How should modern medicine's dramatic new powers to sustain life be employed? How should limited resources be used to extend and improve the quality of life? In this collection, Dan Brock, a distinguished philosopher and bioethicist and coauthor of *Deciding for Others* (Cambridge, 1989), explores the moral issues raised by new ideals of shared decision making between physicians and patients.

The book develops an ethical framework for decisions about life-sustaining treatment and euthanasia, and examines how these life-and-death decisions are transformed in health policy when the focus shifts from what is best for a patient to what is just for all patients.

Professor Brock combines acute philosophical analysis with a deep understanding of the realities of clinical health policy. This is a volume for philosophers concerned with medical ethics, health policy professionals, physicians interested in bioethics, and undergraduates taking courses in biomedical ethics.

Life and Death

Cambridge Studies in Philosophy and Public Policy

GENERAL EDITOR: Douglas MacLean

The purpose of this series is to publish the most innovative and up-to-date research into the values and concepts that underlie major aspects of public policy. Hitherto most research in this field has been empirical. This series is primarily conceptual and normative; that is, it investigates the structure of arguments and the nature of values relevant to the formation, justification, and criticism of public policy. At the same time it is informed by empirical considerations, addressing specific issues, general policy concerns, and the methods of policy analysis and their applications.

The books in the series are inherently interdisciplinary and include anthologies as well as monographs. They are of particular interest to philosophers, political scientists, sociologists, economists, policy analysts, and those involved in public administration and environmental policy.

Other books in the series

Mark Sagoff *The Economy of the Earth*
Henry Shue (ed.) *Nuclear Deterrence and Moral Restraint*
Judith Lichtenberg (ed.) *Democracy and the Mass Media*
William A. Galston *Liberal Purposes*
Elaine Draper *Risky Business*
R. G. Frey and Christopher W. Morris (eds.) *Violence, Terrorism, and Justice*
Ferdinand Schoeman *Privacy and Social Freedom*
Paul B. Thompson *The Ethics of Trade and Aid*

Life and Death

Philosophical essays in biomedical ethics

DAN W. BROCK

BROWN UNIVERSITY

CAMBRIDGE
UNIVERSITY PRESS

Published by the Press Syndicate of the University of Cambridge
The Pitt Building, Trumpington Street, Cambridge CB2 1RP
40 West 20th Street, New York, NY 10011–4211, USA
10 Stamford Road, Oakleigh, Victoria 3166, Australia

First published 1993

Printed in the United States of America

Library of Congress Cataloging-in-Publication Data
Brock, Dan W.
Life and death : philosophical essays in biomedical ethics / by
Dan W. Brock.
p.   cm. – (Cambridge studies in philosophy and public policy)
Includes bibliographical references and index.
ISBN 0-521-41785-6 (hc). – ISBN 0-521-42833-5 (pbk.)
1. Medical ethics. I. Title. II. Series.
R724.B753   1993
174'.24 – dc20   92-2092
                              CIP

A catalog record for this book is available from the British Library.

ISBN 0-521-41785-6 hardback
ISBN 0-521-42833-5 paperback

Contents

v

Contents

PART III: LIFE-AND-DEATH DECISIONS IN HEALTH POLICY

Sources and acknowledgments

These essays were written, each as an independent, self-contained work, over a period of nearly fifteen years, though most are from the latter half of that period. They have been selected from a substantially larger group of essays in biomedical ethics that I have written during this period. In the introduction which follows I will show briefly how the essays are related and state the themes of the larger coherent whole of which they form a part, but I want here to make clear the principles by which they were selected.

While a few of these essays address general issues in moral philosophy concerning taking and saving lives, the large majority are from my work in biomedical ethics; other work in moral philosophy, such as a number of papers on utilitarianism, has been omitted. Even within my work in biomedical ethics, however, I have included my more philosophical essays within the broad theme of life and death, and omitted essays addressed more to practical cases or policy issues. Biomedical ethics is a paradigm of an interdisciplinary field, and my intent here is to provide one example of what the discipline of philosophy has to contribute to its work. The point is not, of course, that there is a clear and clean distinction between the strictly philosophical contributions or contributors to the field and the contributions or contributors from other disciplines. Rather, these essays were written in the belief that many of the underlying philosophical issues, sometimes staples of moral philosophy, have not received enough attention in the biomedical ethics literature in the face of the dramatic pull of individual cases where life and death are at stake. It is my disciplinary belief, and no doubt bias, that our thinking on the dramatic cases and policy issues of life and death that command professional and public attention can be deepened by careful attention to these more philosophical issues. In this respect, by selecting the more philosophical papers, I have at the same time selected what for nonphilosopher readers may be some of my more difficult and less easily accessible work in biomedical ethics. It is left to the reader to judge, of course, whether the more difficult point of entry is warranted by the subsequent philosophical analyses to be encountered. A number of other essays have been omitted because they largely cover ground already relatively easily accessible in book form in my book, coauthored with Allen Buchanan, *Deciding for Others: The Ethics of Surrogate Decision Making*, published in 1989 by Cambridge University Press.

Some readers will recognize a close affinity between some of the views developed in these essays and two of the reports of the President's Commission for the Study of Ethical Problems in Medicine and Biomedical and Behavioral Research, *Making Health Care Decisions* and *Deciding to Forego Life-Sustaining Treatment*. This is not because I have shamelessly borrowed from those reports, but because I served as staff philosopher on the Commission in 1981–82 with special responsibility for some of the more philosophical sections of these two reports. These reports, like the rest of the Commission's work, were truly team efforts and I do acknowledge with gratitude that I did shamelessly accept the good influences of the other very talented members of that staff. Many of the essays in this volume represent my attempts to explore some of the philosophical issues underlying the Commission's work in more detail than was possible in the Commission's reports, though other staff members are not responsible for what I have done with those issues here.

The essays appear here largely as originally published. I have edited a few papers to remove some repetition, but only when doing so did not prevent each essay from standing alone. Since individual essays in a collection of this sort are often read at different times and out of sequence, I have erred on the side of leaving in some repetition in order to retain the individual integrity of each essay. I have resisted the impulse to revise each essay substantially so as to make it say exactly what I now think on each subject in just the form I would now put it, and take solace in the fact that even had I done so, by the time the book appeared in stores such a gap would have reopened in any event.

Sources are listed below. I thank the editors and publishers who gave their permission for the essays to be reprinted here.

Introduction: Previously unpublished.

Chapter 1: "Informed Consent," in D. VanDeVeer and T. Regan, eds., *Health Care Ethics* (Philadelphia: Temple University Press, 1987).

Chapter 2: "The Ideal of Shared Decision Making between Physicians and Patients," *Kennedy Institute Journal of Ethics* 1 (1991): 28–47. Also appeared with minor changes as "Facts and Values in the Physician-Patient Relationship," in E. D. Pellegrino, R. M. Veatch, and J. P. Langan, eds., *Ethics, Trust, and the Professions: Philosophical and Cultural Aspects* (Washington, DC: Georgetown University Press, 1991).

Chapter 3: "When Competent Patients Make Irrational Choices" (coauthor, Steven A. Wartman), *New England Journal of Medicine* 322 (May 31, 1990): 1595–1599.

Chapter 4: "Moral Rights and Permissible Killing," in J. Ladd, ed., *Ethical Issues Relating to Life and Death* (New York: Oxford University Press, 1979).

Chapter 5: "Taking Human Life," *Ethics* 95 (1985): 851–865.

Chapter 6: "Death and Dying," in R. Veatch, ed., *Medical Ethics* (Philadelphia: Jones and Bartlett Publishers, 1989).

Chapter 7: "Forgoing Life-Sustaining Food and Water: Is It Killing?" in J. Lynn, ed., *By No Extraordinary Means: The Choice to Forgo Food and Water* (Bloomington, IN: Indiana University Press, 1986).

Chapter 8: "Voluntary Active Euthanasia," *Hastings Center Report* 22 (March/April 1992): 10–22.

Chapter 9: "The Value of Prolonging Human Life," *Philosophical Studies* 50 (1986): 401–428.

Chapter 10: "Quality of Life Measures in Health Care and Medical Ethics," in A. Sen and M. Nussbaum, eds., *The Quality of Life* (Oxford: Oxford University Press, 1992). Also printed as *Working Paper 66,* World Institute for Development Economics Research, United Nations University, Helsinki, Finland, 1989.

Chapter 11: "The Problem of Low Benefit/High Cost Health Care," printed as a working paper, Center for Health Policy Studies, George Washington University, 1992.

Chapter 12: "Justice and the Severely Demented Elderly," *Journal of Medicine and Philosophy* 13 (1988): 73–99.

Chapter 13: "Justice, Health Care, and the Elderly," *Philosophy and Public Affairs* 18 (1989): 297–312.

Chapter 14: "Truth or Consequences: The Role of Philosophers in Policy-making," *Ethics* 97 (1987): 786–791.

I have been fortunate to have had the good influences of many other scholars in bioethics – through personal and professional contact, as well as from reading their work. To acknowledge them all here would be to list many or even most of the developers of this new field. As philosophical fields go, philosophical bioethics is barely beyond infancy. Most serious philosophical work in bioethics by professional philosophers has been produced within the last three decades, and the great bulk of it within the latter half of that period. Of course, many important contributions to bioethics during this period have been made by nonphilosophers from disciplines such as medicine, law, health policy, and the social sciences. Bioethics has had and continues to have a genuinely interdisciplinary character. Even this larger group of interdisciplinary scholars, however, has been small enough for most of us who have been active in the field to benefit from frequent professional interaction with many or even most other "bioethicists," the name by which we have come to be known.

I want to single out only a few of these bioethicists for

special thanks. I am grateful to my colleague at Brown University, John Ladd, for first drawing me into the field of bioethics many years ago. I mentioned above the singular experience of working with the talented staff of the President's Commission, and I am especially grateful to fellow staff members Alex Capron (the Commission's Executive Director), Joanne Lynn, and Alan Weisbard for their continuing intellectual influences and friendship. The other two "Staff Philosophers" at the Commission, Allen Buchanan and Daniel Wikler, have been for many years, together with Norman Daniels, not only good friends, but also the philosophers with whom I have most often and most profitably discussed many of the ideas contained in the essays in this volume. Dan Wikler also first suggested publishing a collection of some of my biomedical ethics essays, and was extremely helpful in the selection of those that appear here. Allen Buchanan and I have coauthored work which, though it does not appear here, has influenced these essays at many places. I am grateful also to Steve Wartman, who did coauthor one of the essays that is included here, and who also first gave me the opportunity to learn bioethics by teaching the physicians in his residency training program.

Finally, a different sort of thanks to my family, my wife, Chon, and children, Darrell, Dave, and Kate. They gave up much of the time that writing these essays took, and their love and support helped make the work possible and satisfying.

Introduction

A fundamental feature of modern medicine, both for professionals and for the public they serve, is its often life-and-death impact for patients. It is no accident or mere chance that decisions about life-sustaining treatment and care of the dying have been at the center both of the growth of biomedical ethics over the last two decades and of the virtually insatiable public fascination with modern medicine. But the dramatic new powers over life and death that are the great successes of modern medicine have generated widespread concern about how these powers are employed.

During these same last two decades the ideals and norms for physician/patient relations, and to a lesser extent the actual relations between physicians and patients, have been fundamentally transformed. From a largely paternalistic tradition in which treatment decision-making authority was in the hands of physicians and the patients' role was principally to comply with "doctors' orders," a new, more egalitarian ideal of shared decision making between physicians and patients has emerged. From the traditional assumption that a, or even the, central aim of medicine was the preservation of life, a new recognition on the part of physicians and patients exists that medicine's new capacities to extend life are not always a benefit to patients and so must be used selectively.

These two transformations have together brought patients, and so the public, into the process of decision making about life and death. Physicians and patients have together had to rethink their respective roles in decisions which, because life

and death are at stake, are often as intellectually and emotionally difficult and complex as any we face. The essays in
this volume reflect these transformations, were written while
they were taking place, and were and are intended to play
some small part in shaping them.

The essays are grouped into three parts: Part I, Physicians
and Patients Making Treatment Decisions; Part II, Life-and-
death Decisions in the Clinic; and Part III, Life-and-death
Decisions in Health Policy. Part I addresses the transformation that has taken place in the norms and ideals of the
physician/patient relationship and explores the new ideal of
shared decision making. Part II addresses the special moral
issues that arise when treatment decision making is about
life and death. Part III addresses the way these life-and-death
decisions are transformed at the level of health policy, when
the focus shifts from what is best for this patient to what is
just for all patients. In this introduction I want to say a bit
about each essay in order to show how they are related to
each other, to some broader themes, and to the larger bioethics literature.

Part I develops an overall ethical framework for health care
treatment decision making. The first chapter, "Informed
Consent," states the ideal of shared decision making between
physicians and patients and shows how it is served by the
requirement of informed consent. It begins by addressing
the fundamental question of what the aims of medicine, and
in turn medical treatment decision making, should be. A
central issue here for shared decision making is the extent
to which the aims of medicine are defined in objective terms,
for example maximizing health or prolonging life, not dependent on the particular patient's values and goals. I argue
that one major ground for shared decision making is that the
patient's well-being should be the fundamental aim of medicine, and that it has a significant "subjective" component,
in the sense that it depends significantly on the particular
patient's values and goals. Secondly, the important value of
individual self-determination, understood as making significant decisions about one's life for oneself and according to

one's own values, also supports the patient's participation in decision making and ultimate right to decide about treatment. The rest of this chapter elaborates the three standard components of informed consent – that the patient's decision be informed, voluntary, and competent – and some of the main ethical issues in determining when each of these components is adequately satisfied.

My account does not deviate markedly from recent orthodox views of informed consent such as that of the President's Commission for the Study of Ethical Problems in Medicine in its study *Making Health Care Decisions*, though my account of competence for health care treatment decision making does remain controversial. (My account of competence has been developed at considerably greater length in Allen Buchanan's and my book, *Deciding for Others: The Ethics of Surrogate Decision Making*.)[1] Informed consent has been a cornerstone of the new medical ethics that has developed over the last two decades. One major impetus for that development was the reaction to numerous well-publicized ethical abuses in research, most of which in significant part were failures to obtain informed consent.[2] Informed consent too has been the broader basis of support for patients' rights to forgo life-sustaining treatment. It has been one of the central accomplishments of a first generation of work in biomedical ethics to overturn the earlier norm of physician paternalism and to replace it with shared decision making and informed consent. It is now time to look more carefully and critically at the many complexities and qualifications in how that ideal should be understood and applied in relations between real physicians and real patients. Each of the other two chapters in Part I explores important aspects of those qualifications and complexities. I view this as part of the work necessary in a second generation of medical ethics which starts from this new paradigm of shared decision making and informed consent and seeks to make that ideal more sophisticated and thereby more defensible.

Chapter 2, "The Ideal of Shared Decision Making between Physicians and Patients," criticizes what I take to be a relatively common understanding of shared decision making and the roles therein of physicians and patients. I call that understanding the "facts/values division of labor" because it assigns to the physician the provision of facts bearing on the patient's diagnosis and prognosis with various treatment alternatives, including the alternative of no treatment. Physicians call on their training, knowledge, and expertise to provide these facts. This division of labor also assigns to patients the provision of the values and goals to evaluate the different alternative outcomes from different alternative treatments. One inadequacy of this understanding of shared decision making is its failure to acknowledge that physicians too are responsible agents whose moral and professional commitments sometimes will justify an unwillingness to follow patients' wishes, and may require transfer of the patient's care to another physician. But more important, it fails to acknowledge the important role for physicians as advocates for the course of care which in their judgment best serves their patients' well-being. As patients or potential patients, and knowing how our decision making sometimes can be impaired by serious illness, all of us have good reason to want our physicians to act as strong advocates for our well-being, at the same time that they must be willing in the end to respect our self-determination by accepting even our bad but competent treatment choices. The facts/values division of labor between patients and physicians is not defensible, and so my aim in this essay is to present a more complex account of the roles of physicians and patients in decision making about treatment that retains a proper commitment to patient well-being and self-determination, while taking account of the realities of caring for sick patients that physicians too often correctly find missing in the work of bioethicists.

Chapter 3, "When Competent Patients Make Irrational Choices," coauthored with Steven A. Wartman, examines another respect in which shared decision making understood along the lines of the facts/values division of labor is seriously oversimplified. Experienced clinicians know well that patients' treatment choices are often irrational, though not to such a degree that the choices make the patients seem incompetent. Rather than simply accepting an apparently irrational and harmful choice by a patient, a physician has a responsibility, we argue here, to try to understand the source of the irrational choice and to help the patient make a better choice. The different ways in which patients' treatment choices can be irrational have received very little analysis or study. As a first step toward a better understanding, we propose a taxonomy of some of the main forms of irrational treatment choices designed to help make the often difficult and subtle distinction between irrational choices, which the physician should help the patient correct, and merely unusual but rational choices, which should be respected. Each of the last two chapters in Part I poses for physicians the difficult but essential task of defining their role in shared decision making with patients in a manner that recognizes their active and positive responsibilities to help patients work toward sound treatment choices without at the same time slipping back into undue paternalism.

In Part II, "Life-and-death Decisions in the Clinic," I focus on the special philosophical and moral issues that arise for the subset of treatment choices in which the patient's life itself is seriously at stake. The analysis in this part draws on and extends the shared decision-making framework developed in Part I, and addresses the special issues surrounding taking life in clinical contexts. While decisions about life-sustaining treatment must be understood within a broader ethical framework for treatment decisions generally, they must be understood as well within a more general ethical account of the wrongness of killing and of permissible taking of human life. Most work in bioethics on life-and-death decisions fails to spell out and defend adequately, or often at all, the more general moral account of justified and wrongful killing presupposed. The result has been that often the deeper nature of the moral issues in dispute fails to be clarified, distinctions are appealed to which have been shown in the philosophical literature to be, at the least, deeply problematic, and the particular position about life-and-death decisions in medicine is altogether too superficially developed and defended.

Chapter 4, "Moral Rights and Permissible Killing," elaborates the general moral view that persons have a moral right not to be killed, together with the kinds of killing in medicine compatible with recognizing and respecting that right. The paper focuses mainly on the philosophical underpinnings of the life-sustaining treatment debate, and compares the rights view of killing with two prominent alternative views about the wrongness of killing – the consequentialist or utilitarian account that looks to the benefits and harms for all affected by a particular killing, and a duty-based view according to which there is a basic moral duty not to intentionally take innocent human life. I suggest in this essay that the rights account better fits most people's general views about the wrongness of killing and other fundamental features of morality, most significantly the importance of individual self-determination. I focus on two forms of permissible killing in medicine. First, killing can be permissible when the patient has waived his or her right not to be killed; this obviously bears on active euthanasia, but also applies, as I argue in Chapter 6 of this volume, to stopping life-sustaining treatment. Second, killing can be permissible when the right not to be killed is overridden by competing moral considerations; this is especially important for allocation of scarce life-sustaining resources if, as I believe, killing is not, in itself, morally different from allowing to die.

Chapter 5, "Taking Human Life," further develops and defends elements of a rights approach to the general moral issue of killing, comparing it to views, either secular or religious, such as Alan Donagan's that treat killing in terms of a basic moral duty not to intentionally take innocent human life. Understanding the difficulties in this duty-based account of the wrongness of killing is especially important for medicine because of the traditional prominence of that account there. While the discussion in this paper focuses on Donagan's theory, the problems I develop for his view extend to most duty-based theories. There are at least two fundamental difficulties with a duty-based approach to killing like Donagan's. First, duty-based accounts fail to give due weight to the moral importance of consent on the part of the one to be killed, and so in turn give inadequate weight to individual self-determination – one of the central values that guide health care treatment decision making. Second, if, as is often the case, proponents of the duty-based view take the duty to be absolute, then to avoid internal inconsistency as well as implausible implications in certain hard cases, the duty-based theorist is commonly driven to making a morally important distinction between our duty not to kill as opposed to our duty to save or not to allow to die, with the duty not to kill being substantially more stringent. I argue, using Donagan as an example, that this is a mistake and explore and defend the general strategy of argument that others like James Rachels and Michael Tooley have also employed against the view that killing is, in itself, morally different from allowing to die.[3] The nature and implications of that argument are still frequently misunderstood in medical and bioethical quarters. For readers largely unfamiliar with the philosophical literature on the morality of taking life, these first two chapters in Part II should serve both as a brief introduction to that literature and, more important, to show how some of the more abstract issues and arguments in that literature have important implications for moral thinking in medicine. Together, these two essays set out a moral account of the wrongness of killing, through setting out the nature and scope of the right not to be killed, which I go on to apply, sometimes explicitly but more often implicitly, to issues of taking life in medicine in the rest of the essays in Part II.

Chapter 6, "Death and Dying," shifts the focus from more abstract work in philosophy on the morality of taking life to the clinical setting in which decisions about life and death must daily be made. The essay begins with a brief account of the shift from the older heart/lung definition of death to the newer definition which adds the brain death criterion that has now been adopted by all states. The central focus

7

of the essay, however, is the development of what I call an ethical framework for thinking about and making decisions about life-sustaining treatment. The core of that framework is the right of a competent patient, or the surrogate of an incompetent patient, to weigh the benefits and burdens of alternative treatments, according to the patient's values, and to select from among available treatments or to reject further treatment. A defense of this view requires clearing away a number of problematic moral differences which many patients, family members, and health care professionals appeal to in reasoning about life-sustaining treatment. This work was begun in chapter 2 of the President's Commission's report, *Deciding to Forego Life-Sustaining Treatment;* I continue it here at greater length and in more detail than was possible there, and also develop an account of killing and allowing to die in medicine that I failed fully to persuade the Commissioners on the President's Commission to accept.

Most of these moral differences can be thought of as posing possible limitations on the decisions of patients or surrogates and in turn on the actions of physicians. Among the more commonly invoked limitations are: life-sustaining treatment can be withheld or not started in some circumstances in which it must not be withdrawn or stopped; physicians may allow to die, but must not kill; physicians may act foreseeing that an earlier death will result, for example by providing morphine needed to relieve pain, but cannot intend the patient's death; it is permissible to forgo extraordinary or heroic measures, but ordinary treatment must be provided; and forgoing life support is permissible but physician-assisted suicide and euthanasia are forbidden. For each of these commonly invoked moral differences there are two central issues: What is the precise nature of the distinction or difference? Is that difference of moral importance or, more specifically, does it distinguish permissible from impermissible taking of life in medicine? I argue in Chapter 6 that most use of these distinctions in debates and decision making about life-sustaining treatment is unclear or confused on the first issue – that of the precise nature of the difference appealed to –

diff between killing & allowing to die

and is mistaken on the second issue in taking the difference to be morally important in distinguishing permissible from impermissible taking of life. Probably the most controversial feature of the view I defend in this chapter is that much stopping of life-sustaining treatment is killing, not allowing to die, as it is more commonly understood. If this is correct, then it is necessary to acknowledge that morally justified killing of patients is a proper part of medical practice, and so no disputes about permissible actions in medicine can be settled merely by showing that the actions would be killing.

In "Death and Dying" I also discuss briefly the special issue of life-sustaining nutrition and hydration, which many take to be morally different from other forms of life-sustaining treatment. In Chapter 7, "Forgoing Life-Sustaining Food and Water: Is It Killing?," I pursue the issue in more detail by attending to special difficulties in determining whether forgoing food and water is killing. I extend further there the analysis of the difference between killing and allowing to die, develop a different account of that distinction according to which stopping life support is usually allowing to die, and show why I believe that view is mistaken. Some additional important components of the overall ethical framework for life-sustaining treatment decisions are developed here, such as the role of "cause of death" judgments in evaluating the permissibility of physicians' actions.

A number of components of the ethical framework regarding taking life in general and life-sustaining treatment in particular, which are developed in Chapters 4–7 and discussed briefly here, remain quite controversial. The broad structure of that framework, however, has been accepted by many. In particular, that competent patients, or the surrogates of incompetent patients, are entitled to weigh the benefits and burdens of treatment according to the patient's values, and to refuse any treatment judged without benefit or unduly burdensome, has now been widely accepted. In Chapter 8, "Voluntary Active Euthanasia," I turn attention to actions that remain highly controversial among the public and physicians and in public and legal policy. While I con-

9

clude that my assessment of the issues and arguments supports permitting voluntary active euthanasia, my principal
aim, here as elsewhere in the volume, is to clarify the issues
at stake by providing a more careful statement and evaluation
of the arguments than they often receive. I consider this the
main contribution that philosophers have to make to the
interdisciplinary field of bioethics. On the issue of whether,
and for what reasons, voluntary active euthanasia or physician-assisted suicide is wrong, more than on questions
about life-sustaining treatment, I bring the more general account of the morality of killing developed in Chapters 4 and
5 directly to bear. On euthanasia the confusion common in
medicine that killing is never morally justified most directly
leads to an inadequate analysis of the moral issues by im-
plying that no more need be shown than that euthanasia is
killing, which it of course unquestionably is. Instead, I argue
that some killing is morally justified, that the fundamental wrong that killing does is to deny a person a valued
and valuable future, and that, if this is correct, voluntary
euthanasia need not be wrong. Moreover, I go on to argue,
the very same values of individual well-being and self-
determination which underlie the ethical framework for treatment decision making generally, and for life-sustaining
treatment decision making in particular, in some instances provide equally strong support for voluntary active
euthanasia.

Perhaps more than any other issue I take up in this volume,
euthanasia displays the importance of distinguishing the
moral evaluation of individual instances of euthanasia from
the evaluation of whether euthanasia should be permitted
by public and legal policy. Many people who oppose permitting euthanasia would accept my argument that in at least
some individual instances it is morally permissible. They go
on to argue, however, that the likely bad consequences of a
public and legal policy permitting euthanasia sufficiently outweigh any likely good consequences of doing so to warrant
its continued general prohibition. Opponents are correct that

this is where the principal issues lie, but I believe a careful articulation and assessment of the likely policy consequences of permitting voluntary active euthanasia, both bad and good, suggests that the case against euthanasia is sufficiently weak to warrant a limited and controlled experiment with it, such as would follow one or a few states permitting it.

The issue of voluntary active euthanasia illustrates one way in which life-and-death decisions in medicine raise issues of public policy as well as ethical issues for individual clinicians. But this remains public policy concerning decisions and actions by individual physicians and their patients about life and death. Issues of life and death arise for health policy in a very different form in the evaluation of health care delivery and financing programs that determine when and how health care will be used both to prolong lives and to improve the quality of lives. Part III of the collection shifts to this health policy context. Once again, as in Part II, the initial essays address the more general ethical and philosophical issues which succeeding essays then bring to bear more directly on the policy setting. A major difference when attention shifts from the clinic to the health policy arena is the new prominence of issues about allocation of resources. Though it has become increasingly difficult, many clinicians would still prefer to practice without attention to whether the services they provide their patients represent a wise and just use of resources. This is the view of physicians as single-minded and unconstrained advocates for their patients. Few physicians, at least when pressed, would argue that no one should attend to resource use and allocation, but many physicians argue that resource questions should be the concern of nonclinicians, and settled not at the bedside but in the halls of policy. Wherever they are addressed, however, questions remain about how they are to be settled and what contribution philosophical analysis and moral philosophy have to make to their resolution. Aren't these economic issues about which physicians should consult economists, not moral philosophers? Here again, the questions require an interdisciplinary approach, now bringing physicians together with economists and policy analysts, as well as moral philosophers and ethicists.

How much should we spend to save or prolong life, in comparison with using those funds to improve the quality of those or other lives? Few questions of public policy are more perplexing and difficult. While economists are often willing to take the tools they have developed for examining economic questions about the goods and services bought and sold in the marketplace, and employ them in examining these questions of life and death in health policy, the public and health care professionals are often uncomfortable about doing so. They recognize the need to integrate the perspectives of economic and ethical analysis. These are surely in part economic issues, but when and how to use our resources to save some lives, to allow others to die, to promote improvements in the quality of some people's lives but not others, are surely in part ethical issues as well.

Economists have developed methods for measuring the value of the benefits of different social programs generally and health care programs in particular. One typical mode of analysis seeks a measure for allocating limited resources between alternative uses, such as between health care and non-health care programs that prolong lives or improve lives. One of the most prominent of such methods is the so-called willingness to pay approach, which measures how much different individuals would be willing to pay for programs that prolong their lives or reduce risks to their lives. In Chapter 9, "The Value of Prolonging Human Life," I explore some of the ethical assumptions and implications of making individual and social resource allocation decisions by willingness to pay methods. Specifically, I argue that willingness to pay methods do not provide an ethically defensible measure of the overall value or benefits produced by life-prolonging health care programs, but instead measure only the monetary value to an individual with a given income of such programs in comparison with other use of the individual's resources.

Moreover, while the willingness to pay measure takes account of individual differences in preferences for health care as opposed to other goods, it introduces a problematic inequality between rich and poor in the value of health care programs. The other main valuational measure I examine in this chapter, quality-adjusted life-years, or QALYs, provides a measure of benefit that respects equality in a deeper sense by looking only to the effect in prolonging and/or improving the quality of any individual's life.

In Chapter 10, "Quality of Life Measures in Health Care and Medical Ethics," I shift attention from the general policy evaluation tools examined in Chapter 9 to the more focused evaluative measures typically employed within medicine and health policy. What do these evaluative measures tell us, whether implicitly or explicitly, about the concept of the quality of life or of a good life? My assumption is that the value of health care ultimately lies in its contribution to an individual's having a good life in the broadest sense of that term. To evaluate the relative contribution of particular health care provided to an individual patient or supplied more generally as a matter of health policy, it is necessary to be clear on the nature of a good life for persons. My assumption in this essay is that since both physicians and health policy analysts must continually evaluate the effects of health care in promoting a good life for persons, they may be able to contribute to our understanding of the nature of a good life. This is one place among others at which biomedical ethics may help to illuminate more general issues within moral philosophy, instead of the illumination going only in the other direction, from moral philosophy to biomedical ethics. Stephan Toulmin has written about how medical ethics saved moral philosophy, and while there is some hyperbole in that claim, I believe it is correct that sometimes moral philosophy can learn from biomedical ethics, and not always the other way around.[4] There is a gain here for physicians too in greater explicit clarity about the nature of a good life for persons, because at the least one important aim of medicine is to help patients achieve such a life. In a variety of hard cases concerning treatment of individual patients, as well as selection of health policy programs, this clarity can help guide choice.

In Chapter 10 I draw on two main sources for illumination of the notion of quality of life employed in medicine: the ethical frameworks for treatment decision making developed in bioethics and the measures for evaluating different health policy programs developed by health policy analysts. The most striking feature of the former is the importance given to individual self-determination or autonomy of the patient. This is often criticized as an excess of individualistic liberalism which fails to take account of the way in which all of us, all of our lives, live within a variety of groups and communities, from families to larger nation states. But a different view is possible according to which individual self-determination is what Rawls has called a highest-order interest of persons.[5] It reflects a central feature of persons, our capacity not just to engage in purposive, goal-directed behavior, a capacity we share with animals, but to form values and to deliberate about what kind of person we want to become. The capacity to be a self-determining, valuing agent is the fundamental ideal of the person within medical ethics. I show how the focus on function in health policy measures comports with this ideal of the person. The evaluative measures of quality of life within health policy bring out the different forms of primary functions, functional capacities needed in the pursuit of virtually any life plan, to adapt Rawls's notion of primary goods.[6] Finally, these measures also show the need for supplementing "objective" measures of functional capacities with "subjective" measures of personal satisfaction in a full account of a good life.

In Chapter 11, "The Problem of Low Benefit/High Cost Health Care," I use this analysis of the quality of life to explore whether we are now using substantial kinds and amounts of health care whose benefits in extending life and improving the quality of life are insufficient to justify its cost, that is, care that is not costworthy. The incentives of both physicians and patients when making decisions to utilize low benefit / high cost

health care resources, as well as cross-cultural data, suggest that there is substantial use of noncostworthy health care in the United States today. The principal ethical issue raised by this problem is what procedures and criteria should be employed to define costworthy care and to limit our use of noncostworthy care. So long as we retain in the United States our substantially private and extremely heterogeneous health care system, the procedures and places in that system for defining and limiting the use of noncostworthy care will be many and variable, but I propose some general features that those different procedures should satisfy. The precise level of services that is costworthy and that should be available in an adequate level of health care available to all citizens is ethically controversial and not well defined in theories of justice in health care. Democratic decision procedures should be used to determine more precisely that level but they can be informed both by the measures for valuing prolonging life examined in Chapter 9 and by those for evaluating improvements in the quality of life examined in Chapter 10.

One specific area of health care that is often low benefit/ high cost, or buys life-years of low quality, is the treatment of the severely demented elderly. In Chapter 12, "Justice and the Severely Demented Elderly," I identify two paradigmatic approaches to justice in health care: a prudential allocator deciding how to allocate a fair income share to health care over a lifetime and, alternatively, interpersonal distribution determined by appeal to principles of distributive justice. I apply these to the determination of the just claims on the use of social resources for three groups of demented patients: end-stage demented patients in a persistent vegetative state, severely demented but still conscious patients, and moderately demented patients. These claims are importantly affected by issues concerning the nature of persons, as distinguished from human beings, and the nature of personal identity. Patients in a persistent vegetative state, because no longer persons, can no longer be significantly benefited by health care and do not have claims on grounds of justice to social resources for health care; their capacity to benefit from health care is closest to persons who have died. Severely demented but conscious patients have claims to care for comfort, palliation of symptoms, and dignity, but not to life-sustaining care. Moderately demented patients, in the absence of democratically based public decisions to the contrary, have roughly the same claims to care as do nondemented patients.

In Chapter 13, "Justice, Health Care, and the Elderly," I pursue a different, though related, aspect of distributive justice in the allocation of health care resources, the age-group problem – what relevance, if any, age should have on a person's just claims to socially provided health care resources. I use here the occasion of a review essay of Norman Daniels's *Am I My Parents' Keeper?* and Daniel Callahan's *Setting Limits* to explore how theories of intergenerational justice in health care are ultimately imbedded within a broader political philosophy.⁷ What I called the prudential allocator approach in the preceding chapter is employed by Daniels to show that age-rationing is not per se unjust. His account fits within a liberal political philosophy which I argue provides a sound basis for determining just claims between age groups. However, because liberal theories employ a minimal or "thin" account of the good for persons, they have little if anything to say about the ends and meaning of the last stage of life within a full life. Callahan relies on a communitarian political philosophy which requires agreement on a full or "thick" account of the good life in general and the last stage of life in particular. The implausibility of ever gaining general agreement on the proper ends and meaning of the last stage of our lives prevents Callahan's communitarian approach from successfully solving the age-group problem. Although the account Callahan offers of the ends and meaning of aging will not command the consensus necessary for solving the age-group problem, it will be found attractive by many individuals, including liberals, who must decide as individuals what to attempt and hope for in their own old age.

The volume concludes with brief reflections stimulated by

my experience as an academic philosopher having wended my way into the halls of public policy. This final chapter concerns a deep conflict that I believe exists between the roles of academic scholar and of participant in the policy process. These two activities have different goals and norms – crudely put, the goal of academic scholarship is truth, while the goal of policy is to secure policies with the best impacts on citizens' lives. I explore particular manifestations of this conflict when one and the same person finds him- or herself in, and so caught between, both roles.

NOTES

1 Allen E. Buchanan and Dan W. Brock, *Deciding for Others: The Ethics of Surrogate Decision Making* (Cambridge: Cambridge University Press, 1989).
2 Cf. David Rothman, *Strangers at the Bedside: A History of How Law and Bioethics Transformed Medical Decision Making* (New York: Basic Books, 1991).
3 James Rachels, "Active and Passive Euthanasia," *New England Journal of Medicine* 292 (1975): 78–80; Michael Tooley, *Abortion and Infanticide* (Oxford: Oxford University Press, 1983).
4 Stephan Toulmin, "The Tyranny of Principles," *Hastings Center Report* 11 (1981): 31–39.
5 John Rawls, "Kantian Constructivism in Moral Theory," *Journal of Philosophy* 77 (1980): 515–72.
6 John Rawls, *A Theory of Justice* (Cambridge: Harvard University Press, 1971).
7 Norman Daniels, *Am I My Parents' Keeper?* (New York: Oxford University Press, 1988); Daniel Callahan, *Setting Limits* (New York: Simon and Schuster, 1987).

Part I

Physicians and patients making treatment decisions

Chapter 1

Informed consent

The two central aims of the requirement that medical care cannot be given to competent patients without their informed consent are illustrated by the following two cases, in which that requirement is violated.

Case 1. Dr. Smith diagnoses his patient, Mrs. Jones, as having breast cancer. He informs her that a radical mastectomy should be performed as soon as possible, brushes aside her timid attempts to discuss the matter further, and with her silent and grudging acquiescence books her in the hospital for surgery.

Case 2. Mr. Brown is suffering multiple complications and disability, principally from advanced diabetes and renal failure. He has had both legs amputated above the knee and is functionally blind. He is not considered a candidate for a kidney transplant and so faces the ne-

The basic positions, and even some of the language, of this chapter bear a close relation to parts of *Making Health Care Decisions: The Ethical and Legal Implications of Informed Consent in the Patient-Practitioner Relationship*, Volume One: Report, President's Commission for the Study of Ethical Problems in Medicine and Biomedical and Behavioral Research (Washington, D.C.: U.S. Government Printing Office, 1982). I was a member of the commission staff team that drafted that report, with special responsibilities for some of the philosophical issues also discussed here. That report's preparation was truly a team effort and my views on informed consent are deeply indebted to the other principal commission staff members on that project, Alexander M. Capron, Joanne Lynn, Marian Osterweis, and Alan J. Weisbard, though none of them is, of course, responsible for any misuse I may have made of their good influences.

cessity of dialysis treatments for the rest of his life. He is mentally alert, has had his medical situation fully explained to him by his attending physicians, and seems to understand his situation well. Two weeks ago, he decided that he wanted no more painful dialysis treatments, knowing that their termination would quickly lead to his death, and he has remained steadfast in that decision since. Despite his decision, his physician refuses to stop the dialysis treatments, believing that to do so would be tantamount to killing his patient.

The importance of informed consent is different in these two cases, and is perhaps most conspicuous in the absence of consent having been obtained. In Case 1, obtaining valid consent would require informing the patient not only of her medical condition but also of the alternatives available for treating it; ascertaining that the patient is sufficiently competent to understand her situation and make a decision; and then permitting the patient freely to decide about treatment. The doctrine of informed consent is an important aspect of the general norms that structure the physician-patient relationship. It has relevance to the great majority of cases in which decisions must be made about treatment, since it requires a mutual process between physician and patient of informing and discussion (whenever the patient is capable of discussion), thereby leading to a mutually acceptable treatment decision. The doctrine ought to shape in important respects the nature of nearly all encounters and decision making between physicians and patients. It is designed to provide the opportunity for the patient to become more actively involved in the ongoing decision-making process than has often been the case in medicine.

The second case illustrates the special importance of informed consent in a much narrower class of cases. In those cases in which patients and their physicians are unable to agree on a course of treatment, the competent patient has the right to refuse any treatment, even life-sustaining treatment, no matter how strongly the physician or others believe that the treatment should be undertaken. In the crunch, as it might be put, when disagreement between a physician and a competent patient is irresolvable, it is the patient who has final decision-making authority. The authority of the patient to order a particular treatment, however, is more limited than the authority to refuse one. Although physicians should be flexible in making alternatives available to patients, they cannot be compelled to provide treatments that are outside the bounds of acceptable professional practice or that violate their own deeply held moral beliefs. It is the patient's right to refuse any treatment that is embodied in the informed-consent doctrine and that is violated by the physician's forced treatment in Case 2. Because irreconcilable disagreement between physician and patient is relatively rare, this right seldom will be invoked. Nonetheless, the existence of this right, even when not invoked, helps shape the physician-patient relationship. And the focus of the informed-consent doctrine on the ongoing process of physician-patient communication and decision making is not, of course, unrelated to the right to refuse treatment. The ongoing process of informing and discussion would have less value if the patient's decision could be ignored, and the right to refuse a particular (or any) treatment would be less valuable if there were no means of ensuring the availability of information on which such a refusal reasonably might be based. Together, these requirements aim at ensuring that patients can take an active, full role in the ongoing process of decision making about their medical care, and that no treatment is undertaken that does not first meet their approval.

This is not an uncontroversial view of the proper physician-patient relationship and, more specifically, of the role of patients in decisions about their health care. There is a long and strong authoritarian tradition within medicine that denies patients this role. In its more extreme form, this tradition denies any significant patient participation and leaves the ultimate choice of treatment solely with the physician. In this view, patients should be told what is therapeutically useful for them to know about their condition and treatment, and nothing more. Their role is passively to follow the instructions of their physicians. Because medical knowledge and treatment techniques have expanded dramatically in recent years, this traditional authoritarian relationship might seem even more necessary today. Patients simply will lack the knowledge or ability even to understand the information necessary for sound medical decision making. Thus, it might seem only sensible that treatment decisions be left to those with the necessary training, experience, and knowledge to make them soundly – physicians. And especially since patients, who at best understand their situation only inadequately, may make decisions positively harmful to themselves, even to the point of resulting in preventable loss of life. In this view, the physician's role is the paternalistic one of directing treatment in the best interests of the patient, and the patient's role is the largely passive one of following the prescribed treatment.

The informed-consent doctrine amounts to a rejection of this traditional authoritarian and paternalistic conception of the physician-patient relationship. But why should that traditional conception be rejected, given its long and important historical role and its seemingly reasonable character? To answer that question, the values served by the doctrine of informed consent must be examined. Only then can what is at stake in its acceptance or rejection be understood. My concern here will be with the conceptual and moral issues involved in the informed-consent doctrine, and not with the details of its development as a legal doctrine.[1] The focus will be on the *moral* basis and requirements of informed consent.

THE VALUES UNDERLYING INFORMED CONSENT

The informed-consent doctrine serves not only the individual patient in the physician-patient relationship but society as well. Its consequences for the patient are fundamental because the patient is the ultimate recipient of medical treatment. The informed-consent doctrine serves to promote the patient's well-being and to respect his or her self-determination. Both the nature of these values, as well as how the informed-consent doctrine furthers them, must be spelled out.

Patient well-being

Clearly, the fundamental goal of the physician-patient relationship is to protect and promote the patient's health. Medical care treats disease and thereby prevents, eliminates, or ameliorates pain and suffering, disability and premature loss of life. But given the great complexities and sophisticated knowledge involved in such treatment, it is not immediately clear why the patient's health is not best served by vesting ultimate decision-making authority with the physician. Whether a person is healthy or diseased, and what will most effectively treat his or her disease, would seem to be "objective" matters about which well-trained experts such as physicians can best decide. Patients, however, are often confused, anxious, and fearful and as a result sometimes make decisions not in their best interests. They sometimes refuse clearly beneficial treatment, select treatment less efficacious than others available, and fail to complete beneficial treatments. Why then is the patient's well-being best promoted by leaving ultimate decision-making authority with him or her?

The answer lies in three key points: first, and least important, "health" is not a fully "objective" matter, invariant between persons; second, medical criteria alone often do not fully settle which treatment is correct or best for a given medical condition, and third, the relative importance or value of health as compared with other aspects of individual well-being differs for different people.

In perhaps the great majority of cases what constitutes an impairment to health is not controversial – major diseases like cancer or diabetes, fractured limbs, serious infections, and so forth. Because these diseases can have serious adverse effects on an individual's normal functioning and even lead to untimely death, it is largely uncontroversial that they are contrary to a patient's well-being. In a few cases, however, it is debatable whether a condition that may lead to a medical intervention represents any impairment to health at all – for example, an unusually large nose or wrinkled skin for which corrective surgery is possible. However, there is general agreement in the great majority of cases about what constitutes impairment to health. If defining health were the only issue, the case for patient participation in decision making in order to promote the patient's health would be weak.

More important, in much of medicine, it takes more than medical facts alone to determine what treatment is "indicated" for a given condition; there often are several acceptable methods of treatment. A classic example is breast cancer, which can be treated, among other ways, by radical mastectomy, limited mastectomy, lumpectomy, and radiation. Many other conditions are similar. Sometimes the choices vary greatly – for example, between surgical and nonsurgical treatment of a slipped disk. In other cases, the differences between alternatives may be small though still significant – for example, the choice between medications having different side effects for the treatment of severe headaches. Moreover, for many conditions, both treatment and nontreatment are considered acceptable alternatives by medical professionals – for example, concerning hemorrhoids. Whenever more than one acceptable alternative exists, the "medical facts" about which the physician is expert will not determine which treatment is best for a particular patient.

What compounds the difficulty of selecting between alternative treatments is that more than health commonly is at issue. The choice is not simply what will maximize a patient's health, but what will best promote his or her overall well-being. Health, for nearly everyone, is only one value among many, albeit an important one. People value their health in part to avoid the evils of pain and unwanted death. But they value it as well for its prevention of disability and its role in making possible the pursuit of the many other goals and activities that they value. People's values and plans of life vary greatly, as does the importance assigned to health, and particular components of health, in comparison with other goals and values. The physician's ultimate responsibility is to use his or her medical skills to serve patients' overall well-being in this broad sense, to facilitate patients' pursuit of their plans of life. The choice of medical treatment now can be seen to be even less "objective," since it depends on which alternative best fits a patient's overall goals or life plan. For a given medical condition, appropriate treatments will vary greatly because people's goals and values themselves vary greatly. People differ in their tolerance of pain and so, for example, in whether for a slipped disk they prefer surgery with the risks of serious disability or prefer more conservative treatments which may require them to tolerate more pain. An active person may prefer radiation therapy to surgical amputation of a cancerous limb, even if radiation has a higher risk of recurrence of the cancer. Which medical intervention, if any, best will serve a patient's overall well-being must be determined in light of that patient's relevant subjective preferences and overall aims and values. That is what did not happen in Case 1; no attempt was made to determine the importance *to the patient* of minimizing the risk of recurrence of the cancer, of physical disfigurement, and so forth. Following a narrow notion of health in such a case, one that focuses only on minimizing the risk of recurrence of the cancer, may not be in accordance with *the patient's* conception of her well-being and is not the sole proper end of medicine.

It should now be clear why the traditional authoritarian, paternalistic conception of the physician-patient relationship is unacceptable even if the only goal of health care is to promote the patient's well-being. Sound health care decision making involves both objective and subjective components, and both must be taken into account. On the objective side, there are empirical facts about the nature of the expected outcomes for the patient of different treatments (including the alternative of no treatment). These include a determination of the patient's physical status, including his or her prognosis and expected condition under alternative treatments. These are factual or objective matters about which the patient's physician typically will be the best judge. There may often be significant uncertainty concerning these consequences, however, and medical experts will frequently disagree about such factual matters.

The subjective component of decision making involves assessing the relative importance or value to this particular patient of the features and consequences of the alternative treatment outcomes. This evaluation of the outcomes, and of their effects on the patient's well-being, is a subjective component in the sense that it depends on the patient's goals and values. These values and goals may be quite different from the physician's, or from most people's, depending as they do on the patient's preferences, abilities, and opportunities. The authoritarian and paternalistic tradition in medicine fails adequately to recognize the relevance and importance of these subjective components. Because the assessment of the effects on a patient's well-being of treatment alternatives requires a meshing of these objective and subjective components, a process of mutual interaction or shared decision making is called for, with the physician and patient each contributing what each is in the best position to know.

Patient self-determination

The nature of self-determination. The other principal value promoted by the informed-consent doctrine is patient self-determination. A person's interest in self-determination reflects the common desire to make important decisions about one's life oneself and according to one's own aims and values. Self-determination involves the capacities of individuals to form, revise over time, and pursue a plan of life or conception of their good. It is a broad concept applicable at both the levels of decision and of action. Patients' exercise of self-determination in the context of health care involves their deciding which alternative treatment will best promote their own particular goals and values. As I will discuss later, this requires that they become informed about the nature and consequences of alternative treatments. More relevant to the nature of self-determination, it requires as well that patients have formed a minimally stable and consistent conception of their good – that is, a set of goals and values, or plan of life, by which to evaluate those alternatives. Having a conception of one's good is more than merely having a set of desires, instinctual or conditioned. People have a capacity for reflective self-evaluation, for considering what kinds of desires and character they want to have. There are, of course, limits to the extent to which people can change their desires. They can change their particular tastes in food, but not whether they have any desire to eat at all. In broad respects, our natures are fixed and given to us by our biological nature, but within these broad limits we adopt particular values and create a unique self. In these ways, we are capable of shaping our character and of taking responsibility for the kinds of people we are. Self-determination is the name given here to that process.

To say that we are capable of self-determination is neither to deny our interdependence nor that our values are influenced by others. Still, in the evaluation of these influences, one either makes them one's own and incorporates them into one's own conception of what is good, or rejects them. Self-determination does *not* imply that persons create their own character and values free of all causal influence from others. Nor does it require free will in the sense of a will free of causal determination.

The formation of one's values or conceptions of what is good is an ongoing process. In the light of new experience, people revise their values over time. Our interest in taking responsibility for our life by forming a plan of life extends as well to being free to revise and develop our values.

People not only can form and revise a conception of their own good but also can pursue it in action (e.g., in pursuing a particular medical treatment). Thus, interference with self-determination can involve interference with people's deciding for themselves, but also interference with their acting as they have decided they want to act. Respecting self-determination consequently involves avoiding both these interferences, with decision and with action.

Self-determination thus embodies a particular ideal of the person. That ideal is one of a person freely choosing for him- or herself, on the basis of personal, reflectively adopted, values, the alternative action to be taken, and then so acting. The spirit of this ideal has been eloquently expressed by Isaiah Berlin:

> I wish my life and decision to depend on myself, not on external forces of whatever kind. I wish to be the instrument of my own, not of other men's acts of will. I wish to be a subject, not an object; to be moved by reasons, by conscious purposes, which are my own, not by causes which affect me, as it were, from outside. I wish to be somebody, not anybody; a doer – deciding not being decided for, self-directed and not acted upon by external nature or by other men. . . . I wish, above all, to be conscious of myself as a thinking, willing, active being, bearing responsibility for his choices and able to explain them by reference to his own ideas and purposes.[2]

The value of self-determination. What weight should be given to patient self-determination in health care decision making? This question is particularly pressing in those cases in which a patient has decided on a treatment course seemingly contrary to his or her own well-being. Refusals of life-sustaining therapy are some of the most dramatic examples. In Case 2 above, it may be plausible to suppose that continued treatment would not promote the patient's well-being. But if the patient had not been disabled, was capable of living a rewarding and satisfying life with long-term dialysis treatment, and *then* refused it, respecting his self-determination would appear to require a great sacrifice of his well-being. Evaluating the importance of the impact of medical treatment on a patient's life plan provides a framework for weighing the relative value of medical care for his or her well-being. It is less obvious what the value of self-determination resides in, and so what weight it should be given.

Self-determination typically is valued both for the good consequences that result from respecting it (its instrumental value) and for its own sake, as part of an ideal of the person (its noninstrumental or intrinsic value). Perhaps the most important instrumental value of respecting self-determination lies in its role in promoting a person's good or well-being. This may seem paradoxical, since in discussions both of informed consent and of paternalism it is common to pose individual well-being and self-determination as competing and conflicting values. Indeed, these values sometimes are in conflict, but their relationship is more complex than is suggested by viewing them simply as opposing values. I have argued that a person's well-being or good is largely subjective in that it is largely determined by that person's values and subjective preferences. This is the idea behind the view that the individual is the best judge of his or her own interests. Respecting a person's self-determination leaves a person free to pursue and attain those values and goals. In the exercise of self-determination in purposive, rational action, we pursue our good as we perceive it. An important value of self-determination, and in turn an important reason for respecting it, is then that its exercise commonly results in the promotion of the person's good or well-being. This is not to say that a person can never be mistaken about where his or her good lies. But, when realized, the ideal of self-determination (understood to involve the adoption and pursuit of a plan of life) itself usually contributes to the realization of a person's good or well-being. That this is an important part of the value of self-determination can be seen by the fact that when people's choices bear little relation to their good, for example in cases of severe and wide-ranging psychotic delusion, the extent to which their exercise of self-determination is of any value at all, and certainly of any intrinsic value, is decidedly problematic. Part of the value of respecting people's self-determination then lies in the way in which doing so *normally* contributes to the promotion of their good. This has implications for those instances when persons are confused or ill-informed and so make decisions contrary to their own good. In those cases, not only must protecting their well-being be weighed against respecting their self-determination, but also in that weighing the value of self-determination itself is diminished in comparison with its value in more favorable circumstances.

Although the value of self-determination is then diminished in such cases, it is not lost. That is because respecting self-determination often has other instrumental values such as the avoidance of frustration involved in interfering with a person's liberty of action, the development of individual judgment (especially since people often learn best from their mistakes), the satisfaction people often get from making decisions about their life for themselves, and so forth. The extent to which these instrumental benefits are present in any particular case depends on the facts of that case.

The instrumental consequences of self-determination are not, of course, all good ones. Sometimes persons make choices that are very harmful to themselves or others, experience anxiety from choosing for themselves, and so prefer to have others decide for them. The overall instrumental effects of any particular exercise of self-determination must be assessed in context.

The value that many people ascribe to self-determination rests also on a particular noninstrumental ideal of the person – we value being, and being recognized by others as, the kind of person who is capable of determining and taking responsibility for his or her destiny. This is the ideal of the person expressed in the quote from Berlin above. It is a *noninstrumental ideal* in the sense that it offers an attractive vision of what human beings can be, independent of the consequences or satisfactions in realizing that vision. There is a dignity in being self-determining that is lost in even a satisfying subservience. It is this ideal of the person that is expressed in patients' desires to make significant decisions about their lives for themselves, even if others (for example, physicians or computers) might be able to predict how they would decide and so decide for them, or even if others could decide in a way that would more effectively promote their health and well-being.

Waiving self-determination. Patients often do not take an active role in decision making about their own health care. Often they say to their physicians something like, "You do whatever you think is best." This may seem contrary to the account of self-determination as involving people making significant decisions about their lives for themselves. In turn, it might be thought not to satisfy the doctrine of informed consent. *Must* patients ultimately make health care decisions themselves, or may they either waive the exercise of self-determination in particular instances or, perhaps better, exercise it at a different level of decision? The moral doctrine of informed consent as interpreted here *entitles*, but *does not require*, a patient to take an active role in decision making regarding treatment. Patients sometimes have good reasons for avoiding an active role in particular decisions about their health care – for example, they are gravely ill and want to spend what little time remains to them with their families and others, or they know that thinking about the issue at hand makes them distraught and depressed, or they know from past experience that the decisions require technical understanding of which they are incapable and that trying to decide only makes them confused. In any of these circumstances, and especially if a patient has a physician who knows him or her well and who is trusted to act in the patient's best interests, it could be reasonable to let the physician decide. In these cases, the patient waives his or her right to decide and transfers that right to another – his or her physician or, alternatively, a family member. Doing so is compatible with, and a possible use of, the right to self-determination. What is necessary is that the patient understand that the decision is rightfully his or hers to make, and so also his or hers to transfer to another if the patient so chooses.

By contrast, other patients may leave decisions with their physician because they do not think they have any business interfering in the physician's professional decision about treatment – it never occurs to them that the decision is theirs to make. This would be one plausible rendering of the first case. And this could not possibly be an exercise of self-determination, because a person must first believe that he or she has a right to decide before he or she could intend to transfer it to another. But when a patient does freely and intentionally transfer the right to decide to another, this should be understood as compatible with the right of self-determination and with informed consent. Nevertheless, two qualifications are necessary.

One ideal of self-determination is the self-governing person who makes decisions for him- or herself, after personally weighing alternatives against reflectively adopted values. In turning over a major health care decision to another, a patient does *not* meet that ideal of self-determination. But even that ideal is a matter of degree and not fully realizable – for example, we all rely on much information and accumulated knowledge that we never could evaluate for ourselves. The shortages of time and ability (among other factors) that make such reliance reasonable can also sometimes make turning a health care decision over to another a reasonable and legitimate use of one's right to decide.

The second qualification is that this is one place among others where the moral analysis of informed consent may diverge from the doctrine's optimal legal form. The law and legal policy must be formulated in order to limit well-intentioned misuse and ill-intentioned abuse of its requirements. We want general rules that, when applied by real persons in a variety of circumstances, will produce the best results on balance and over a period of time. The worry here is that permitting true and valid waivings of the right to decide would inevitably bring in its wake other denials of that right, along with failures adequately to inform and involve patients who would want to decide for themselves. It may be that a legal exception to the requirements of informed consent to permit such waivings should be narrowly limited because of the other unavoidable and unjustified denials of patients' rights to decide that it would produce. Whether that is so depends on the particular procedural safeguards that might be devised to permit such waivings while limiting abuses, and empirical investigation of their actual effectiveness.

Other values served by informed consent

In the account of informed consent being developed here, the principal benefits secured by informed consent are benefits for the patient whose consent must be obtained – his or her well-being and self-determination. However, undoubtedly other values are served by informed consent, especially when it is viewed as an institutionalized social practice. It fosters general public trust in the medical enterprise, especially in the context of medical research in which serious concerns would otherwise exist about conflicts of interest between medical researchers and subjects. The requirement that physicians explain and justify treatment recommendations to the patient may also encourage more careful scrutiny and review of those recommendations by the medical profession, thereby resulting in sounder recommendations. There are many other effects of the complex institutional practice of informed consent. Some of those effects are often claimed to be negative; for example, patients may lose some of their trust and faith in the physician's healing powers when all the risks and uncertainties of treatment must be detailed to them. An overall assessment of these institutional benefits and costs is far from clear, and probably too uncertain to serve as the fundamental moral basis of the informed-consent doctrine.

Although the values of patient well-being and self-determination may provide the underlying moral basis for the doctrine, it is helpful to examine the requirements and limitations of the doctrine to see how they do so. As I have already noted, the values of patient well-being and self-determination may be in conflict when a patient's decision is contrary to his or her well-being. (Sometimes, as in the case of a comatose patient, a patient may be unable to make any decision at all.) Only when it is clear how the informed-consent doctrine deals with such conflicts will we know whether it adequately secures patient well-being and self-determination.

THE CONDITIONS FOR VALID INFORMED CONSENT

The requirements of a morally valid consent are commonly considered to be three: that the person giving consent be *competent;* that he or she be *informed* about the proposed intervention; and that his or her consent be *voluntary*. Competence is necessary to ensure that a patient has sufficient decision-making capacities to participate in the decision process and to decide responsibly on a course of treatment. The requirement that the person be informed is to ensure that the patient has received all relevant information concerning available alternative treatments (including no treatment), including the risks and potential benefits of each, in order to be able to judge which alternative is most desirable. The voluntariness condition ensures that the patient's choice does not result from coercion, duress, undue influence or manipulation, and thus will serve the patient's own perceived good rather than another's good, or another's view of the patient's good. Each of these conditions is complex, both in principle and in practice, and I will examine each in turn. But it is important to emphasize that together these conditions *do* ensure that the patient's consent to treatment is both an exercise of self-determination and a decision reasonably expected to promote his or her well-being.

Competence

The requirement that a patient be competent in order to give binding consent is more complex than first appearances might suggest. In some cases it is clear whether the patient is competent. Some people are unable to participate in decision making about their health care to any significant extent

36

at all – the comatose, infants and very young children, and some of those who are severely mentally disabled or mentally ill. With these groups, little or no possibility exists of involving them in decision making about their own health care. Nevertheless, most "normal" adults of average intelligence are usually able to understand their situation and decide about their treatment. In between are the difficult cases – those involving people whose capacities to decide are marginal and whose ultimate competence to decide requires clarity about the nature and function of the determination of competence.

The importance of competence. The importance of the determination of competence can be put simply – it separates those patients whose free and informed decisions must be accepted as binding from those whose decisions will be set aside. In other words, the competence determination ultimately establishes when a patient's consent for treatment *must* be secured and honored (i.e., in which cases the requirement of informed consent will be operative). And this suggests that the basic values of patient well-being and self-determination underlying the informed-consent doctrine itself will be at stake in the competence determination, as indeed they are. Consider the issue in personal terms. Suppose you are concerned both for your own well-being and your self-determination. When would you want your own decision about your treatment to be accepted and followed, and when would you want others empowered to decide for you? If your decision-making capacities are sound and are being effectively utilized, your own decision should reflect your view of your well-being – both values will be promoted by your deciding. However, if your decision making is seriously defective, the likelihood increases that your decision may cause you serious harm or the loss of important benefits, and in general not conform to your values. In those circumstances, it is reasonable to want protection against one's choices being honored. A normal concern for our own well-being requires it. The value of self-determination may be diminished in such circumstances; nevertheless, the values of deciding for oneself and protecting one's own well-being are then in conflict. The determination of competence must secure an acceptable balance between allowing persons to decide for themselves while protecting them from the harmful consequences of their own choices when their capacities to decide are seriously limited or defective.

Several points follow from this moral and legal role of competence judgments. First, competence is always relative to a particular task – everyone is competent to do some things and not others. The relevant task here is to make a particular health care treatment decision. It follows that a person may be competent to make a given treatment decision, but incompetent to manage his or her financial affairs or even to make other treatment decisions. Second, patient decision-making competence also changes over time, for example, from the effects of medications or in cases of a cyclic mental illness like manic depression. Health care professionals can and should take steps as necessary to enhance patient competence. Third, though the capacities contributing to competence are possessed in varying degrees, the determination of a person as either competent or incompetent is an all or nothing judgment. The crucial question will then be *how* defective a person's decision making must be in a particular instance to be deemed incompetent. To understand how to answer that question, we must examine more closely the various elements of decision-making competence.

Elements and standards of decision-making competence. In order to participate in health care decision making and ultimately to give valid consent to treatment, an individual must be able to communicate and understand relevant information, be able to reason and deliberate about alternative treatments, and possess values and goals by which to assess alternatives. The nature of the requirements that a patient be able to understand information and reason about alternatives will be detailed below in discussion of the condition that consent be informed. Understanding and reasoning are together necessary for a patient to be able to acquire and use information to reach an appreciation of the consequences of adopting one or another course of treatment. The patient's values and goals provide the basis for evaluating which course of treatment will best serve his or her ends.

Individuals possess these capacities, and exercise them on particular occasions, in different degrees; or, what amounts to the same thing, the capacities to understand, reason, and value can be imperfect or defective in more or less extensive and important ways. Patients' decision-making competence, as expressed in the making of a particular treatment decision, thus could be measured, at least on a relative scale. While no single, uniform measure of decision-making competence seems possible, this merely reflects the reality that competence involves a complex meshing of a patient's various capacities and skills with the demands of the particular decision situation. The potential situations are too numerous, and the potential arrays of patient abilities and inabilities are too varied to permit a single measure. But from the lack of a formal measure of competence it does not follow, as is often charged, that determinations of competence and incompetence lack any theoretical foundation and are merely "intuitive" in the sense of being inchoate and arbitrary.

This attack on competence determinations as at base intuitive and arbitrary has seemed to some to be reinforced by the widely varying standards or levels of competence actually applied by physicians, courts, and others. However, those differences do not support the "intuitive and arbitrary" charge. Instead they in part reflect confusions and moral disagreements regarding the proper level of competence required and on the issue of paternalism, and in part reflect a quite proper variation in the standard of competence. Standards of competence are divided below into three broad classes, moving from the most minimal to the most stringent standard, in that the alternatives will classify a progressively larger proportion of people as incompetent. While the minimal standard is insufficiently paternalistic, the strictest standard is overly paternalistic. The proper standard lies in the middle category and evaluates the *process* of the patient's decision making, though even this standard should vary depending on the decision in question.

Ability to express a preference. This standard of competence considers only whether the patient is able to express a preference for a particular treatment but does not assess the presence or nature of the understanding or reasoning underlying that preference, nor the content of the preference itself. It maximally respects patient self-choice or self-determination, but it provides no protection at all against the consequences of defective decision making. For example, this standard would require accepting an individual's refusal of life-saving therapy when that refusal was caused by psychotic delusions induced by drugs or mental illness. This standard is in fact not a standard of *competent* choice at all, since any choice whatsoever is deemed competent. It fails to provide any protection of patient well-being – one of the two values that the informed-consent doctrine is designed to promote. It fails to balance the values of self-determination and well-being, but instead defers entirely to the former, even under conditions in which the value of self-determination is drastically diminished.

Standards evaluating the content of the patient's decision. While the "mere expression of a preference" standard of competence is insufficiently paternalistic, at the other extreme are standards that determine competence by evaluating the content of the patient's choice and that are excessively and objectionably paternalistic. This is not to deny that it is commonly disagreement with the content of the physician's recommendation that will trigger review of the patient's competence in marginal cases. This is natural and reasonable, given the presumption that the physician's recommendation is in the patient's interests. But it is to say that merely having decided on a course of treatment different from the physician's (or family's) recommendation serves as no basis at all for a determination of incompetence. Basing determinations of incompetence on the content of the patient's decision inevitably requires appeal to some "objective" standard of correct decisions, independent of whether that standard conforms to this patient's values and goals. "Correct" decisions in this view might be those that are rational, or what most would decide, or what the physician would decide, and so forth. But it is highly problematic whether any such objective standard of a patient's good is ultimately defensible. Employing such a standard will seriously compromise the patient's self-determination, without requiring evidence that his capacity for self-determination in the circumstances is deficient. Finding a patient incompetent merely because his treatment choice does not conform to some external, supposedly objective, standard fails to respect the patient's capacity to define his own values. Finally, any such standard in practice risks substantial abuse and unwarranted denial ("for their own good") of patients' rights to decide for themselves.

Standards evaluating the nature of the patient's understanding and reasoning. Any assessment of competence that examines a patient's actual understanding and use of information in reasoning to reach a decision allows protection against serious harms that may result from defective understanding and mistaken reasoning. This standard of competence focuses on the *process* of decision making, rather than either on merely whether the patient has made a choice or on merely the content of his or her choice. Of course, the first response of health care professionals or others to apparent misunderstandings or mistakes of patients should be further discussion with the patient to seek to remove or remedy the difficulties. Only when such misunderstandings or mistakes prove irremediable is there any warrant for a finding of incompetence. The goal under this standard of competence, as with the informed-consent doctrine generally, is to foster the patient's full participation in decision making that most clearly conforms with his or her own values and goals. This standard seeks a balance between respecting a patient's self-determination and protecting his or her well-being. When

41

information about the nature of available alternatives and their consequences for the patient's values and goals has been understood and appreciated by the patient, then his or her choice must be honored, even if others such as the physician would have chosen differently. But when the patient is clearly unable to understand necessary information and make a reasoned decision, a standard that focuses on the nature of the patient's decision-making *process* allows a possible finding of incompetence which would justify transferring the decision to a surrogate.

While the general standard of competence should focus on the process of the patient's decision making, this still leaves open two important questions regarding the competence assessment. What level and degree of understanding and appreciation of the full consequences of alternatives are necessary for competence? How certain must others be that the patient has not achieved the appropriate level before finding him or her incompetent? The answer to both questions is, "It depends." The level of decision-making incapacity that should be required for a finding of incompetence properly varies depending on the decision in question. There is *no* single proper level of decision-making capacity that should be required for all treatment decisions. This is simply a consequence of the fact that what is a reasonable balance between patient self-determination and well-being in setting that level will depend in part on the consequences of the patient's choice for his well-being.

The consequences of acting on a patient's choice can range along a continuum from being substantially better to being substantially worse than other alternatives in their expected effects in achieving the goals of preserving life, preventing injury and disability, and relieving suffering, as against their risks of harm. The relevant comparison should be with other available alternatives, and the degree to which the net benefit/risk balance of the alternative chosen appears better or worse than that for other treatment options. When the net benefit/risk balance appears substantially to favor the patient's choice it is reasonable to require only a low/minimal level of decision-making capacity in the choice; there is no need to compromise his self-determination in order to protect his well-being. At the other extreme, when the expected effects for the patient's life and health appear to be substantially worse than available alternatives, a requirement of a high/maximal level of competence is reasonably required to ensure that the patient's well-being is adequately protected and that in exercising his self-determination he has chosen in accordance with his own aims and values. When the expected effects for life and health of the patient's choice are approximately comparable to those of alternatives, a moderate/median level of competence is appropriate to ensure that the patient's well-being is adequately protected while respecting his self-determination.

One final and perhaps unexpected consequence of this variability is that the level of competence for consenting to and refusing a particular treatment need not be the same, because the consequences for the patient's well-being may be dramatically different. It is reasonable to insist that a patient fully understand that he or she is *refusing* a life-saving therapy that has only limited risks, while applying less strict standards to the patient's *acceptance* of that same therapy. While this variability in morally appropriate levels of competence implies the need for discretion on the part of individuals making competence determinations, legal and professional practices must be designed to limit the unjustified use of that discretion wrongly to deny patients' rights ultimately to insist on what they believe will best promote their values and goals. The more reason we have to distrust professional use of such discretion, the more we will reasonably design legal rules and practices to limit it.

Voluntariness

For the consent of a competent decision maker to be binding, it must be voluntary or freely given. Consent that is obtained as a result of coercion or duress, manipulation or undue influence, is invalid and does not authorize treatment. An involuntary choice will not reflect the aims of the chooser, but the will of the coercer or manipulator instead, and so will violate the chooser's self-determination.

Coercion. A patient's decision is coerced when the patient is threatened, either explicitly or implicitly, with unwanted and avoidable consequences unless the patient makes the desired choice. For example, a gravely ill patient is told that unless he agrees to further treatment, other palliative measures to limit his discomfort will be withdrawn. Outright coercion of patients, at least by health care professionals, is probably relatively rare, though it may be more common in special settings, such as mental hospitals, prisons, and some nursing homes. The great inequalities between involuntarily committed mental patients, prisoners, children, some nursing home populations, and those on whom they are dependent, make these groups especially vulnerable to coercion. Since there is significant potential for conflict of interest between these groups and their caretakers, special scrutiny of the voluntariness of their consent to medical care is warranted. Coercion by family members to force treatment is probably more common than by health care professionals. With relatively few exceptions, such as inoculations for public health purposes or treatment in connection with obtaining evidence in criminal proceedings, coerced or forced treatment of competent patients is not morally justified or legally permitted. Forced treatment without the patient's consent can be distinguished from coerced consent. Forced treatment involves no consent, while coerced consent invalidates the consent; each makes the treatment unauthorized.

It is sometimes said that patients faced with very difficult and unpleasant alternatives, such as whether to undergo painful chemotherapy when it is the only treatment for an otherwise fatal cancer, do not choose the chemotherapy voluntarily because their choice is coerced by the disease itself. Nevertheless, unpleasant and unwanted choices between bad alternatives are not in themselves coercive. When the undesired alternatives are caused, for example, by the patient's illness, rather than by the threat of another person, they represent the unfortunate but inevitable constraints within which the patient must decide about treatment. That all alternatives are "bad" and leave little or no "real choice" provides no sound reason to set aside the patient's choice as involuntary and to transfer the decision to another.

It is also important to distinguish coercion involving threats of unwanted consequences from warnings about the natural history of a disease and the natural consequences of certain decisions. In most cases, the distinction turns on whether the physician acts to bring about the unwanted consequence (threat) or merely informs the patient that it will be a consequence of one of the patient's alternatives (warning). Telling a hypertensive patient of the medical consequences of not taking medication constitutes a warning and does not coerce him or her to take the medication, whereas threatening to withdraw further care unless the patient accepts the physician's preferred treatment generally is coercive. The first is part of a physician's responsibility, while the second is morally wrong.

Manipulation. Although coercion is probably relatively uncommon in health care relationships, manipulation of patient decision making is more widespread. Indeed, it is a common criticism by physicians that the informed-consent process is meaningless because they can get the patient to agree to whatever they want. Manipulation in health care relationships is complex, both in theory and practice, and can take more forms than there is space to discuss here. Many features of the health care setting make it ripe for manipulation. Patients are commonly worried and fearful, uncertain of what is "wrong" with them, and seek reassurance and care. They are explicitly asked to place their trust in physicians who possess special knowledge and capacities to give care. In these circumstances, the disparities in knowledge, status, and authority make manipulation sometimes hard to avoid. Perhaps the most blatant form of manipulation is when information is deliberately withheld from patients in order to affect their choice. For example, they are not told of alternative treatments to the one the physician prefers, or of significant risks or side effects of a recommended treatment. This clearly limits a patient's capacity to make an informed choice that best fits his or her particular needs and values.

Manipulation can, however, be far more subtle than outright deception. For example, a physician's manner and tone of voice can indicate whether a risk is one to worry about, or whether a concern is justified. The way information is presented can affect how the patient responds to it – for example, "This treatment succeeds in two out of every five cases" or "This treatment fails most of the time." In some cases, there may be no fully "objective" or "neutral" form in which necessary information can be conveyed. Nevertheless, it generally is possible to avoid putting information in a form designed to change the patient's decision from what it would have been with a sound understanding of the information. Such manipulation may be conceived as being "for the patient's own good" and producing the "best" decision. Or it may be used merely to avoid the more difficult and time-consuming process of helping the patient clearly understand the situation and how alternative courses of action may or may not serve his aims and needs. But in the former case it both infringes self-determination by denying patients a sound understanding on which to base their choice and fails to respect the extent to which people's own aims and values determine what will best promote their own well-being. In the latter case, although the outcome of patients' decisions may not be different as a result of the manipulation, the manipulation bypasses patients' own deliberative processes and their self-determination is compromised, making the decision not fully theirs.

None of this is to say that physicians should not make treatment recommendations to patients who want those recommendations, or that such recommendations are inherently manipulative. On the contrary, it is part of physicians' professional responsibility to make such recommendations available, and these recommendations naturally and reasonably influence patients' decisions. If the recommendations are tailored to a particular patient's goals and values and are not manipulative in the ways discussed above, they need not compromise the voluntariness of the patient's decision.

The various forms of involuntariness of treatment decisions occur in different degrees – coercion and manipulation may both be more or less serious, irresistible, and pervasive in undermining self-determination. Ideal physician-patient decision making would involve no kind or degree of involuntariness. The law and other forms of public policy should seek to promote this ideal. Nevertheless, legal remedies in the law for coerced or manipulated consent often require evidence both that (1) the patient's decision was different from what it would otherwise have been as a result of the involuntariness and that (2) the treatment given resulted in harm to the patient. This is only one instance of the general fact that the law is neither a desirable nor effective instrument for preventing or remedying every moral wrong, or every failure to attain a moral ideal.

Informed understanding

The goal of the requirement that consent be informed is for patients to achieve a sufficient understanding of their condition and possible treatments so that they can make a sound assessment of which treatment, if any, will best serve their goals and values. It thereby permits an informed exercise of self-determination and promotes a decision most in accord with the patient's well-being. It is worth adding as an important practical matter that therapeutic benefits can accrue from the patient's being well-informed (for example, increased conformity to a treatment regimen), that the great majority of patients want vital information about their treatment,[3] and that the lack of an adequate explanation of the nature and purpose of a procedure is a major cause of refusals of treatment by patients.[4] Physicians sometimes conceive of informed consent as principally requiring a disclosure of the risks of treatment in order to protect themselves from legal liability should a problem arise. This leads to association of informed consent with the form that the patient signs testifying that such risks have been disclosed. But such forms are neither necessary nor sufficient to protect from legal liability, and although obtaining informed consent can protect against legal liability for battery (though not from liability for malpractice), that is not its main ethical basis.

The focus on *disclosure* of risks to protect against legal liability also has had the unfortunate result of shifting attention away from the *comprehension* or *understanding* of the information by the patient. A patient's understanding is the result sought by any requirement that a patient be informed. Thus, physicians should tailor the presentation of information to the particular patient so that it is as comprehensible as possible. This usually will involve avoiding technical medical or scientific jargon and helping the patient understand the consequences of various medical procedures. Making information comprehensible to the particular patient also is relevant to meeting the objection often made by physicians that informed consent cannot be achieved because patients only rarely can fully understand the relevant information. In its extreme form, this objection holds that patients would have to be given extensive medical and scientific training if they are to understand fully their medical condition and various treatment alternatives. Such objections, however, are largely misguided. Physicians obviously bring medical knowledge to the treatment decision process that patients nearly always lack, just as patients bring knowledge of their own particular goals and values that physicians lack. It is not necessary for a patient to become a physician in order to give informed consent. In order to make an informed decision about treatment, patients do not require all the medical details and underlying scientific bases of treatment alternatives that their physicians have. Rather, patients need to understand how those treatment alternatives will affect their capacity to pursue their various plans of life. This understanding rarely will require a sophisticated medical or scientific background.

The other aspect of this objection concerns the impossibility or unreasonableness of communicating all information possibly relevant to the decision at hand. Even if it all could be communicated, such information would overwhelm the average patient who is unable to integrate it all into his or her decision process. Moreover, telling patients of every risk or possible side effect of treatment, no matter how remote, might so terrify them that they would reject needed treatment. This objection raises questions about the *kinds* and *amounts* of information that ought to be conveyed to the patient. The broad kinds of information to be conveyed are relatively uncontroversial. They should include: (1) the patient's current medical condition, including a future prognosis if no treatment is pursued; (2) any treatment alternatives that might improve the patient's condition and prognosis, including an explanation of the procedures involved, the significant risks and benefits of the alternatives, with their associated probabilities, and the financial costs of the alternatives; and (3) a recommendation as to the best alternative. We assume for the present that the patient has not expressed a desire *not* to have any of this information. If there is anything not commonly discussed in this list, it is the financial cost of the alternatives. However, when the costs to the patient are significant (and whether they are will vary with a patient's circumstances), they can have a substantial impact on the patient's financial ability to pursue other goals important to him or her – and so on the overall desirability of treatment alternatives, and can affect the likelihood of the patient's actually pursuing a treatment course.

What is controversial is the amount of information that must be conveyed. More specifically, concern has focused on when risks are too remote or inconsequential to require that the patient be told about them. Courts, as well as the state legislatures that have intervened on this matter, have largely been divided between two alternative general standards of disclosure: (1) the standard of professional medical practice for such cases – crudely put, an individual physician must convey particular information, and in particular disclose risks, if most other physicians do or would do so in similar cases; (2) the standard of what a reasonable person would want to know in the case at hand – or, as it is sometimes put, whatever information is "material" to the decision at hand. The first standard probably provides a clearer guide for physicians, as well as greater protection from liability for nondisclosure. However, given the long paternalistic tradition of limited disclosure of information to patients, the "professional practice" standard fails to ensure that patients will have the information necessary for sound decision making. The "reasonable person" standard, though introducing greater uncertainty about just what information must be conveyed, is more likely to result in patients obtaining necessary information. However, as a moral matter, this standard should be amended to require conveying whatever information a reasonable person would want to know, *plus* whatever else the actual person wants to know. For example, a professional athlete may quite reasonably want to know about a risk from treatment that would ruin her career that others would not want disclosed to them. So it is important that physicians consider the particular concerns and values of the actual patient and how these may differ from other persons. Perhaps the simplest way to try to ensure meeting any unusual information needs a particular patient may have is always to invite the patient to ask for any additional information he or she may want after a discussion of information the physician considers material to the decision.

Legal policy

Here, as elsewhere in this essay, the focus is on the ethically sound basis and nature of informed consent. There may be good reason not to fully individualize the *legal* requirement for disclosure because of the difficulty of establishing in litigation, often taking place long after the event in question, what the particular patient wanted to know. Doing so would create an invitation for self-serving testimony by patients, for whom a remote risk has eventuated in actual harm, that "of course, I would have wanted to know about that." For similar reasons, it may be best for the law not to allow *less* disclosure than a reasonable person would want, even though fully respecting self-determination would support doing so when patients do not want to be told that information. As a matter of public and legal policy, it is probably best not to fully individualize the information requirement, in order to help protect against the abuse of patients being denied information they do or would in fact want.

The requirement that a patient be informed before consenting often is claimed to be limited by a physician's "therapeutic privilege." Specifically, it is claimed that physicians are justified in not informing competent patients of particular risks when doing so would seriously upset the patient. Therapeutic privilege has the potential of seriously undermining the informed-consent doctrine, given that physicians are quite naturally averse to conveying unpleasant information to their patients, and that patients quite naturally become upset at learning it. Consequently, any ethically acceptable exception to the requirement of informed consent based on therapeutic privilege must be carefully and narrowly framed. Rather than merely "upsetting" the patient, it should be the case that disclosing the information would cause serious harm to the patient. And the harm must result from the receipt of the information itself, independent of its causing the patient to make a different treatment decision than he or she would otherwise have made. Moreover, there should be strict evidentiary requirements for the harm, such as clear and convincing evidence of the high probability of imminent and serious harm. Any such standard will require much interpretation in individual cases – when is evidence clear and convincing, what probability is high enough, when is harm imminent and serious? It would permit temporarily withholding from a deeply depressed and suicidal patient devastating information highly likely to trigger a suicide attempt. However, in the interest of limiting its abuse such a standard would rightly set a high burden of proof on physicians who sought to invoke therapeutic privilege. It is probably reasonable as well to require, when therapeutic privilege is invoked and important information is withheld from a competent patient, that the procedures required for incompetent patients involving use of a surrogate decision maker be followed.

Therapeutic privilege is one of the exceptions to the general requirement that treatment cannot be given without the patient's informed consent. I have discussed in detail a second exception – incompetence. I have also discussed a third exception – waiver by patients of their right to decide. The final exception to the informed-consent requirement is emergency – treatment can be given without a patient's consent when failure to render urgent care would likely result in serious harm or loss of life to the patient, and obtaining the patient's consent is either not possible or would seriously delay the rendering of that care.

The overall argument of this essay can be summarized in a sentence: giving due weight to the value of personal well-being while respecting individual self-determination morally requires gaining and honoring the informed and voluntary consent of competent patients for their medical treatment. However, such a simple summary should not be allowed to obscure two points. First, even the more detailed analysis of informed consent here is not detailed enough. It leaves many issues unaddressed – for example, what role, if any, should patients' unconscious beliefs and motivations play in determinations of their decision-making competence, voluntariness, and understanding? And it leaves many cases hard cases – for example, in instances in which a patient's well-being and self-determination are in conflict, we lack any precise and uncontroversial basis for assigning a weight to each. But even the best and most detailed analysis leaves genuinely hard cases hard and controversial, though it should help us understand better why they are hard and what conflicting values and arguments make them controversial. Second, the moral ideal of informed consent expressed here is a high ideal, which is often realized imperfectly at best in medical practice. There is not space here to explore the many avenues of change that might help move much medical practice closer to that ideal. But there can be little doubt that fully realizing the ideal would involve significant transformations of the traditional health care process.

NOTES

1 Some of the important legal cases bearing on informed consent include Schloendorff v. Society of New York Hospital, 211 N.Y. 125, 105 N.E. 92, 95 (1914), in which the right to self-determination is asserted; Salgo v. Leland Stanford Jr. University Board of Trustees, 154 Cal. App. 2d 560, 317 P.2d 170 (1957), in which "informed" was added to the consent requirement to yield "informed consent"; Natanson v. Kline 186 Kan. 393, 350 P.2d 1093, the opinion on denial of motion for rehearing, 187 Kan. 186 354 P.2d 670 (1960); and Canterbury v. Spence, 464 F.2d 772 (D.C. Cir), cert. denied, 409 U.S. 1064 (1972) and Cobbs v. Grant, 8 Cal. 3d 229 104 Cal. Rptr. 505, 502 P.2d 1 (1972), both of which enunciate standards of disclosure of information.
2 Isaiah Berlin, "Two Concepts of Liberty," in *Four Essays on Liberty* (Oxford: Clarendon Press, 1969), 118–138.
3 See the survey of public and professional attitudes and practices in the area of informed consent performed for the President's Commission by Louis Harris Associates, in *Making Health Care Decisions: The Ethical and Legal Implications of Informed Consent in the Patient-Practitioner Relationship*, Volume Two: Appendices, President's Commission for the Study of Ethical Problems in Medicine and Biomedical and Behavioral Research (Washington, D.C.: U.S. Government Printing Office, 1982).
4 See the study of treatment refusal by Paul Applebaum and Loren Roth in *Making Health Care Decisions*, Volume Two, appendices.

SUGGESTIONS FOR FURTHER READING

Capron, Alexander. "Informed Consent in Catastrophic Disease Research and Treatment." *University of Pennsylvania Law Review* 123 (December 1974): 340–438.
Childress, James F. *Who Should Decide? Paternalism in Health Care.* New York: Oxford University Press, 1983.

Drane, James. "The Many Faces of Competency." *Hastings Center Report* 15 (April 1985): 17–21.

Dworkin, Gerald. "Autonomy and Informed Consent." *In Making Health Care Decisions: The Ethical and Legal Implications of Informed Consent in the Patient-Practitioner Relationship*, Volume Three: Appendices, Studies on the Foundations of Informed Consent. President's Commission for the Study of Ethical Problems in Medicine and Biomedical and Behavioral Research. Washington, D.C.: U.S. Government Printing Office, 1982.

Freedman, Benjamin. "A Moral Theory of Informed Consent." *Hastings Center Report* 5 (August 1975): 32–39.

Kass, Leon. "The End of Medicine and the Pursuit of Health." In *Toward a More Natural Science: Biology and Human Affairs.* New York: Free Press, 1985.

Katz, Jay. *The Silent World of Doctor and Patient.* New York: Free Press, 1983.

Lidz, Charles W., et al. *Informed Consent: A Study of Decisionmaking in Psychiatry.* New York: The Guilford Press, 1984.

Making Health Care Decisions: The Ethical and Legal Implications of Informed Consent in the Patient-Practitioner Relationship, Volumes One through Three. President's Commission for the Study of Ethical Problems in Medicine and Biomedical and Behavioral Research. Washington, D.C.: U.S. Government Printing Office, 1982.

Meisel, Alan; Roth, Loren H.; and Lidz, Charles W. "Toward a Model of the Legal Doctrine of Informed Consent." *American Journal of Psychiatry* 134 (1977): 285–289.

Roth, Loren; Meisel, Alan; and Lidz, Charles W. "Tests of Competency to Consent to Treatment." *American Journal of Psychiatry* 134 (1977): 279–284.

Tancredi, Laurence R. "The Right to Refuse Psychiatric Treatment: Some Legal and Ethical Considerations." *Journal of Health Politics, Policy and Law* 5 (Fall 1980): 514–522.

Veatch, Robert M. "Three Theories of Informed Consent: Philosophical Foundations and Policy Implications." In Appendix B to *The Belmont Report: Ethical Principles and Guidelines for the Protection of Human Subjects of Research*, Volume Two, chapter 26, pp. 1–66. DHEW Publication No. (OS) 78–0014. Washington, D.C.: U.S. Government Printing Office, 1978.

Chapter 2

The ideal of shared decision making between physicians and patients

INTRODUCTION

Shared treatment decision making, with its division of labor between physician and patient, is a common ideal in medical ethics for the physician-patient relationship.[1] Most simply put, the physician's role is to use his or her training, knowledge, and experience to provide the patient with facts about the diagnosis and about the prognoses without treatment and with alternative treatments. The patient's role in this division of labor is to provide the values – his or her own conception of the good – with which to evaluate these alternatives, and to select the one that is best for himself or herself. As a rough guide to practice, this is a reasonable conception; most of the time it is likely to produce sound treatment decisions. However, as an ideal it is too simplistic, and is subject to several challenges that I will explore in this essay.

Some challenges relate to the *physician's* role. This facts/values division of labor seems to assume that the physician can and should provide the facts about treatment alternatives in a value-neutral form. But some have questioned whether the sciences on which medicine is based are, or can be, value-

Adapted from "Facts and Values in the Physician-Patient Relationship," by Dan W. Brock, in *Ethics, Trust, and the Professions: Philosophical and Cultural Aspects*, Edmund D. Pellegrino, Robert M. Veatch, and John Langan, eds. Washington, D.C.: Georgetown University Press, 1991. Reprinted with permission of the publisher.

free. Moreover, the concepts of health and disease, and of the normal and pathological, are held by many to be value-laden. And even if it is possible for physicians to provide only value-neutral facts to the patient, should the physicians' role be restricted in this way?

Other challenges are to the *patient's* role as provider of the values for the evaluation of the different treatment alternatives. One defense of this role is based on the claim that physicians are not in a position to know reliably what is in a patient's best interests (Buchanan 1991). I believe at least as influential – even if often not explicitly stated – is an extreme subjectivism about values that assumes that the patient's own ultimate values, which define his or her own conception of the good, are incorrigible. By incorrigible, I mean they cannot be mistaken. In the standard case in which treatment is pursued for the patient's benefit, this extreme subjectivism is thought to support the patient's values guiding treatment decision making and treatment. I will explore this incorrigibility claim and will argue that on each of the main philosophical conceptions of the good for persons it is indefensible and should not be the basis for shared treatment decision making.

Underlying this view of the physician-patient division of labor are assumptions and beliefs about the nature and relation of facts and values, and science and ethics, that are largely the legacy of a logical positivism that has long since been rejected by most philosophers. It is time that these unstable foundations for our normative ideal of the physician-patient relation were finally removed and replaced with more defensible underpinnings.

TWO MODELS OF THE PHYSICIAN-PATIENT RELATIONSHIP

Positivists insisted on a relatively sharp distinction between descriptive or empirical claims and science, on the one hand, and evaluative claims and ethics, on the other.[2] Descriptive statements and the claims of science were to be true or false according to whether they in fact correctly described and explained the world. By remaining properly descriptive or empirical, science could, at least in principle, be entirely value-neutral or value-free.

In its more extreme versions, positivism held that ethical judgments lacked cognitive content, but were instead expressions of emotions or attitudes, and as such could be neither true nor false, correct or mistaken. Because no evaluative statements were held to be logically entailed by any descriptive or empirical statements, moral reasoning was thought to be properly understood not as reasoning but as attempts at nonrational persuasion.[3] Many drew from this an extreme subjectivism in ethics along the lines of what I have called above the incorrigibility thesis – there is no meaningful sense in which an individual's ultimate values defining his or her own conception of the good life could be mistaken, false, or unfounded.

These positivist views – and in particular this extreme subjectivism about values – need at the very least substantial qualification and revision. They did, however, contribute to one extremely important and beneficial historical change in medicine, specifically in ideals of the physician-patient relation. Until roughly the last two decades, the physician-patient relationship in this century could be characterized as authoritarian or paternalistic. Moreover, this was the generally accepted normative ideal for that relationship, not just a reality that failed to measure up to the ideal. Physicians, because of their medical training, knowledge, and experience, were seen as the experts on what treatment would be best for a particular patient, and treatment decisions were thus appropriately left in their hands. The patients' role was largely passive – to follow "Doctor's orders" – and so patients needed only the information necessary to ensure their compliance with treatment, not information to enable them to participate in treatment decisions.

The positivists' insistence on the distinctions between facts and values, and science and ethics, contributed to the eventual erosion of the physician's claim to a virtual monopoly on treatment decision-making expertise and authority. Positivism thus contributed significantly to a substantial reduction in unjustified paternalistic action by physicians on behalf of patients.[4]

A few critics of the paternalist model of physician-patient relations would turn it on its head, giving patients sole decision-making authority. The physician's role would be to carry out the patient's (or surrogate's, in the case of incompetent patients) orders for treatment. One author labeled this the "physician as body mechanic," the physician being the one to whom we bring ourselves in for whatever repairs we desire (Bayles 1978). Robert Veatch (1972) some years ago labeled this the engineer model of the physician's role – the physician is the technical expert who knows what can be done to or for the patient and who is trained to carry out the various procedures. The physician's job is to provide just what the patient orders in this revision of "Doctor's orders."

This ideal of physician-patient relations has few defenders, though the positivist legacy provides it with more support than is often thought. Two main considerations have been paramount in the rejection of this consumer sovereignty model. First, it fails to take account of the reality of sick patients who, suffering the physical and psychological effects of their illness, are often fearful, anxious, and deeply desirous of putting their treatment in the hands of health care professionals they can trust (Sherlock 1986). Many patients' ordinary decision-making capacities are substantially diminished by serious illness. And even when a patient's ordinary decision-making capacities remain intact, the enormous expansion of medical science, knowledge, training, technology, and treatment capacity would seem to leave the average patient ill-equipped to make decisions about his or her own health care. Second, this consumer sovereignty model fails to acknowledge physicians as independent agents with professional and moral commitments that limit the extent to which they should simply respond to and serve patients' wishes or desires.

As noted earlier, the model of ideal physician-patient re-

lations that has largely replaced the paternalistic or authoritarian model is often characterized as "shared decision
making." In this approach, both physicians and patients have
essential roles in ideal treatment decision making. Physicians
are to use their knowledge, training, and expertise to provide
their patients with a diagnosis and a prognosis if no treatment
is undertaken, together with information about alternative
treatments that might improve the prognosis, including the
risks and benefits and attendant uncertainties of such treatments. Patients articulate their own aims, preferences, and
values in order to evaluate which alternative is best for them.
The idea here is that, on one hand, uninformed patient
choices or consent will lead to decisions that fail to best serve
patient interests – thus, the need for physician participation.
On the other hand, decisions guided by physicians' values
will fail to reflect patients' self-determination interest in making important decisons about their lives for themselves –
thus, the need for patient participation.

Although shared decision making is an ultimately sound
ideal, a simple physician-patient division of labor between
facts and values cannot be plausibly maintained. One difficulty is that it appears ultimately still to leave physicians'
actions fully under the direction of patients. No more than
the consumer sovereignty model does it recognize physicians
as independent agents, with moral and professional commitments that can appropriately guide and limit their action
in the service of a patient's aims and values. A second difficulty is the assumption that, in principle, facts and values
can always be distinguished, and in practice can be sufficiently distinguished to permit the physician-patient division
of labor.

A third difficulty with the facts/values division of labor is
that it assumes that the patient's values or account of the
good is correct and so must be accepted and respected by
the physician, whatever its content. This is due in part to
acceptance of what I have called the extreme subjectivity of
values – if a patient's conception of his or her own good
cannot be false or mistaken, and cannot be logically incom-

patible with any facts, then there is no basis for rejecting them as incorrect, false, or unfounded. (Nor is there any basis for accepting them as correct, true, or well founded, a point commonly overlooked by extreme subjectivists.) Of course, it will be acknowledged that patients can be, and often are, mistaken about factual matters concerning their good in a number of ways. They can be mistaken about what will or will not, in fact, lead to some valued state of affairs – for example, which treatment will produce a desired or valued outcome. They can be mistaken about how they may adjust to a particular treatment outcome over time, and how they may come to find satisfaction in new activities and pursuits. But what patients cannot be mistaken about – what is incorrigible in this account – are the ultimate values that define their conception of the good or their life plan.

THE INCORRIGIBILITY THESIS AND THE PATIENT AS VALUES-PROVIDER

Incorrigibility vs. self-determination

Some will quite rightly argue that a reason other than incorrigibility of values underlies physicians' acceptance of patients' values or conceptions of their good in shared treatment decision making. In fact, the most common explicit reason offered for accepting patient choices about treatment, as reflected in the doctrine of informed consent both within the law and medical ethics, is that doing so respects patients' self-determination or autonomy. Self-determination is not only important when patients are choosing wisely; it also protects at least some bad or defective choices. The right to self-determination includes at least some right to make mistakes in choosing our own life path, and so it can be the importance of patients' self-determination, not the incorrigibility of their values and conceptions of their own good, that undergirds the facts/values division of labor and respect for patients' choices in shared treatment decision making. This appeal to self-determination is correct in the sense that persons commonly do value making important decisions about their lives for themselves.

This is true even in some circumstances in which they may grant that others might make better decisions on their behalf, even as judged by their own values. Making such decisions for ourselves is a way in which, for better or worse, we take responsibility for and control over our lives.

The importance of self-determination requires that physicians nearly always respect the voluntary and informed treatment choices of competent patients. But a fundamental distinction must be made: whether competent patients can be mistaken in their conception of their own good and in the ultimate values that define that conception is distinct from whether physicians or others should paternalistically interfere with or set aside a patient's treatment choices if and when they are mistaken. The importance of self-determination bears only on the latter issue. And even if personal self-determination should nearly always bar paternalistic interference, it is important for the ideal of shared decision making whether the extreme subjectivity of values – with its implication of the incorrigibility of patients' ultimate values – is correct. If it is, then besides providing information about diagnoses and prognoses with alternative treatments, the physician's role in shared decision making would seem at most to be to seek to discern what the patient's values are (thus, the talk in some quarters that replaces moral reasoning with "values clarification") so as to help ensure that treatment choices serve those values.

This still seems closer to the consumer sovereignty model than to a model of shared decision making that acknowledges the physician as an independent moral and professional agent. Failure to differentiate between the issue of the corrigibility of patients' own conception of their good, and the issue of whether the importance of self-determination largely bars paternalistic interference by physicians, can also weaken resistance against unwarranted paternalism. The quite reasonable rejection by physicians of patients' incorrigibility about their own good can mistakenly lead physicians to infer that paternalistic interference with their patients' competent decisions about treatment is thereby warranted.

The champions of individual self-determination, then, are at least correct that it provides an alternative basis for a presumption for ultimate deference to patients' choices in shared treatment decision making. But we still need to examine more directly and critically the claim that a person's ultimate values and conception of the good are incorrigible. I believe this claim should not be accepted, however natural and even obvious it might seem to many.

Incorrigibility based on metaethical subjectivism

Two versions of this extreme subjectivity of values need to be distinguished and discussed separately – what I will call metaethical subjectivity and normative subjectivity. First, I take up the issue of metaethical subjectivism.

Until the recent revival of interest among philosophers in moral realism, much metaethical concern focused on the problem of the justification of moral judgments (Sayre-McCord 1988; Brink 1989). If epistemological intuitionism in ethics of the sort associated with Moore, Prichard, and Ross had proved defensible, then the problem of justification would have been "solved" by our faculty of ethical intuition together with the truth of the ethical beliefs that it guaranteed. For a variety of reasons, well known at least to philosophers, intuitionism and the ethical truth it was supposed to certify have been widely rejected. The alternative of ethical naturalism, too, came under a variety of widespread criticisms in the 1950s and 1960s because it seemed to leave out the evaluative content of ethical discourse, and because appeal to naturalistic definitions seemed too slim a reed on which to rest the entire enterprise of justification in ethics.[5] Both ethical intuitionism and naturalism are cognitivist accounts of ethical judgments according to which those judgments have truth values that are usually explicated in terms of correspondence theories of meaning and truth.

The current conventional philosophical view of justification in ethics acknowledges that ethical judgments do express attitudes, as the positivists had it, but also that they have substantial cognitive content and are backed by reasons that make them capable of support and admitting of reasoned argument. In this view, justification is usually developed along coherentist lines with one or another version of John Rawls's "reflective equilibrium" (1971, sec. 9) at the heart of the view. Coherentists hold that a moral judgment is justified if it survives a critical screening process for reaching wide reflective equilibrium among all of a person's moral judgments.

In coherence theories of the justification of patients' evaluative judgments about their own well-being or good, these judgments are subjective in the sense that they are justified only to the extent that they survive the process of reaching reflective equilibrium by that person. That process, however, is a complex one, the details of which cannot be explored here. Suffice it to say that there are many places in that process at which a person's initial moral judgments may undergo revision and correction. No plausible coherentist view gives any reason for physicians to accept as incorrigible patients' judgments about their good, or even the ultimate values by which patients define their good. Incorrigibility is never achievable for moral judgments since the best that individuals can hope for is to reach wide reflective equilibrium in their moral judgments at a particular point in time (Daniels 1979). There is no assurance that new experiences, considerations, or arguments may not in the future upset that equilibrium and cause some of the moral judgments it encompasses to be revised.

Coherence accounts of justification in morality do make justified moral judgments ultimately subjective in the sense that the moral judgments that are justified for a particular person are those that person makes in reflective equilibrium. Since reflective equilibrium is only an ideal at which a person can aim rather than a state one could be assured of being in before considering a significant ethical choice, there is never reason to assume at the outset of shared decision making that the relevant values a patient brings to that decision making are maximally justified for him or her in having survived reflective equilibrium.

Incorrigibility based on normative subjectivism

It is important to distinguish this metaethical subjectivism concerning the epistemic status of a person's judgments about his or her good or well-being from normative subjectivism about those judgments. By normative subjectivism of moral or value judgments about the good for persons, I mean the view that what is valuable or good for a person consists entirely in or depends importantly on contingent psychological facts about that person, such as what makes him or her happy or what he or she desires.[6] Normative subjectivism about value is also thought to provide support for the facts/ values division of labor in shared decision making. However, there are at least two serious difficulties in this use of normative subjectivism about value or the good. First, even leading subjective conceptions of individual good or well-being do not make a patient's evaluation of treatment alternatives incorrigible. Second, the most plausible theory of value or the good for persons is not fully subjective. Both of these difficulties can be brought out by considering briefly the three broad alternative theories commonly distinguished in the philosophical literature of the good for persons – what I will call hedonist, preference satisfaction, and ideal theories of a good life.[7] Hedonist and preference theories are normatively subjective; ideal theories contain normatively objective components.

Hedonist theories of the good for persons. What is common to hedonist theories is that they take the ultimate good for persons to be certain kinds of conscious experiences. The particular kinds of conscious experiences are variously characterized as pleasure, happiness, or the satisfaction or enjoyment that typically accompanies the successful pursuit of our aims and desires. Particular states of the person that do not make reference to conscious experience, such as having diseased or healthy lungs, and particular activities of the person, such as studying philosophy or playing tennis, are part of a good life in this view only to the extent that they produce the valuable conscious experience. Hedonist views have been widely rejected for failing in serious ways to match most persons' considered moral judgments about a good life (Griffin 1986, chap. 1). Nevertheless, even if a subjective hedonist view of the good were correct, it would not justify an uncritical acceptance of the values a patient brings to shared decision making. Persons often are mistaken about what will bring them pleasure or make them happy; others are often in a better position to determine this for them. No incorrigibility assumption about the patient's evaluation of treatment alternatives is justified on hedonist theories of the good.

Preference satisfaction theories of the good for persons. Preference satisfaction theories take a good life for persons to consist of the satisfaction of their desires or preferences. Desires or preferences have objects and are satisfied when their objects are realized; for example, my desire to see the sunrise tomorrow is satisfied only if I see the sunrise tomorrow.

This is to be distinguished from any feelings of satisfaction or pleasure, understood as a conscious experience of mine, that I may experience if I see the sunrise tomorrow. The difference is clearest when the object of my desire is not an experience. In such a case, my desire can be satisfied, but I do not or could not know it and so I receive no feelings of pleasure or satisfaction from getting what I desire. Preference theories share a subjectivity with hedonist theories since they make a person's good depend ultimately on what that particular individual desires or prefers.

Nevertheless, just as with hedonist theories, preference theories do not support any assumption of patient incorrigibility in the evaluation of treatment alternatives. In this theory, what is good for persons is what they desire for its own sake (what I call their basic desires) as opposed to what they desire only as instrumental, or as means to the satis-faction of their basic desires. It might seem that patients' basic desires are incorrigible, though, of course, patients can be mistaken about the factual question of what courses of treatment will best satisfy those basic desires. But a person's basic desires would incorrigibly determine his or her good on a preference theory only if that theory defined a person's good solely by that person's *actual* basic desires. No defender of the preference theory, however, defends it in this form.

For preference satisfaction theories of the good for persons to be at all plausible, they must allow for some correcting or "laundering" of a person's actual preferences (Goodin 1986). The most obvious example is the need to correct for misin-formed preferences; for example, a person's desire for a par-ticular treatment based on a false belief about his or her medical condition. But apparently basic desires or prefer-ences will sometimes require correction as well; for example, a preference for avoiding medical care because of a fright-ening experience in a hospital as a young child. Corrected preference theories of the good for persons are difficult to evaluate decisively because what corrections should be made to actual preferences is controversial. Nevertheless, the need for corrections of preferences makes clear that even the basic desires or preferences patients express in treatment decision making have no claim to incorrigibility in defining their good until they have either been corrected or it has been deter-mined that they need no correction.

Ideal theories of the good for persons. We have seen that even in normatively subjective hedonist and preference theories of the good for persons, no presumption in shared decision making that patients' values are incorrigible is warranted. The second difficulty with the appeal to hedonist or pref-erence theories to justify support of normative subjectivity and the facts/values division of labor, is that neither a he-donist nor a preference theory is the most plausible theory of the good for persons.

Ideal theories hold that at least part of a good life consists neither of any conscious experience of a broadly hedonist sort nor of the satisfaction of the person's corrected preferences or desires, but instead consists of the realization of specific ideals of the person, including possession of certain virtues and character traits. For example, many have held that an important component of a good life consists in being a self-determined or autonomous agent, and that this is part of a good life for a person even if he or she is neither happier as a result nor desires to be autonomous.

Ideal theories will differ both in the specific ideals, virtues, and character traits the theories endorse, and in the relative place they give to happiness and preference satisfaction in their full account of the good for persons. Ideal theories are normatively objective, or at least contain objective components, in the sense that they hold a good life to be, at least in part, objectively determined by the correct or justified ideals of the good life, independent of the psychological states of the person.

The most plausible theory about the good for persons, and the theory that best fits the considered moral judgments of most members of our society, is an ideal theory, although there is insufficient space to develop, much less defend, these claims here. However, I want to include enough about this conception of the good to make at least its broad outlines clear and to show its relation to the incorrigibility thesis and the division of labor assignment of values to the patient.

One fundamental ideal is the exercise of a person's capacity as a valuing agent, what I have called self-determination. This ideal is central in a good life in the sense that it organizes and determines many other important components of a theory of the good for persons. The exercise of self-determination in constructing a relatively full human life will require in an individual what I have called elsewhere primary functional capacities (Chapter 10 in this volume). By primary functions I mean human functions that are necessary for, or at least valuable in, the pursuit of nearly all relatively full and complete human life plans and lives.

There are four broad types of primary functions: biologic

(e.g. well-functioning organs); physical (e.g. mobility); mental (e.g. a variety of reasoning and emotional capabilities); and social (e.g. the ability to communicate). They have no sharp boundaries and can be specified in different ways. The important point for my purposes here is that in an ideal theory, these functions are normatively objective components of a good life. Their value does not depend on contingent psychological facts about a particular person, although their relative weight within any particular life may be in part subjectively determined in this way.

Besides primary functions, there are what I call agent-specific functions, which are necessary for a person to pursue successfully the particular purposes and life plan he or she has chosen. This could mean, for example, the capacity to do highly abstract reasoning required in mathematics or the physical dexterity necessary to a musician. The relative value of these functions can vary widely among persons and will depend on the particular life plan chosen by the person in question. Because of their role in making possible a significant range of opportunities and alternatives for choice, primary and agent-specific functions are both necessary in a theory of the good life that gives centrality to the valuing and choosing agent.

At an even more agent-relative level are the particular desires pursued by particular persons on particular occasions in their valued aims and activities. These will be more subjective still in their dependence on the specific aims of a particular person. As one moves from primary functions to agent-specific functions to agent-specific desires, one simultaneously moves across a continuum of objectivity to subjectivity in the normative content of a person's good. Also at the subjective end is the final important component of the good for persons – the hedonistic or happiness component of a good life: that aspect which represents a person's subjective, conscious response, in terms of enjoyment and satisfaction, to the life he or she has chosen and the activities and achievements it contains. The happiness component is subjective both because different things make different persons happy and also because reasonable persons can and do disagree about the relative importance of happiness in a good life.

The importance of what I have called the objective-subjective continuum in a full ideal theory of the good for persons is that it explains why the incorrigibility thesis regarding the patient's values is mistaken at the objective end of the continuum and why that thesis becomes increasingly more plausible as one moves to the subjective end of that continuum. The more the patient's values and choices in shared decision making appear to be in conflict with his or her objective good, that is, ideals and functions at the objective end of the continuum, the stronger the case for the physician being an advocate for those ideals and functions and seeking to ensure that the patient's values and choices do not, in fact, conflict with them. This is not to say that the physician should ride roughshod over a patient's values and choices if they appear to be in conflict with his or her objective good. As already noted, the importance of self-determination for a competent patient militates strongly against doing that. However, it is to say that the incorrigibility thesis about the patient's ultimate values is mistaken and cannot justify any sharp division of labor between physician and patient along facts/values lines.

If physicians and patients are exploring alternative treatments when the selection of treatment is genuinely problematic, it is a mistake to believe that the only appropriate role for the physician is the delivery of facts. On the contrary, the physician can have a responsibility to explore, together with the patient, the values by which alternatives should be evaluated and, if necessary, to help in a joint process of critical reflection on those values.

Proponents of shared treatment decision making should not defend it with appeal to the incorrigibility of the patient's values. Physicians are rightly skeptical of that appeal and it can unnecessarily undermine, instead of strengthen, shared decision making by resting it on unsound foundations. It is personal self-determination, not incorrigibility regarding the ultimate values that define one's good, that justifies deference to competent patients' choices in shared decision making.

THE PHYSICIAN AS FACTS-PROVIDER

The other side of the division of labor in shared decision making assigns to the physician the role of gatherer and presenter of facts regarding the patient's diagnosis and prognosis with alternative treatments. The aim is to place the physician in a value-neutral role so as to leave the evaluation of alternatives to the patient according to the patient's values. One central reason for this is the extreme subjectivism about values or, in other words, the incorrigibility thesis. I have argued that that thesis, in both its meta-ethical and normative ethical versions, is mistaken. As a result, physicians sometimes should play an active role in a collaborative process with patients in examining and assessing the values that will guide treatment choice. Nevertheless, what of the role that the division of labor assigns to the physician as a provider of the facts about diagnosis and prognosis? Can that role be value-neutral? Or, instead, does it, or must it, carry value and ethical commitments?

Several versions of the claim of value neutrality for the physician's role should be distinguished, each of which raises different issues. The first version concerns whether *in practice* it is possible for physicians to provide to patients the information relevant to treatment decision making in a value-neutral form. There are a number of ways in which physicians can and do fail to achieve this value neutrality in practice.

First, the language they use to characterize treatment alternatives may not be solely descriptive, but carry implicit or explicit evaluations as well – for example, when a family is told that continued treatment will only prolong the patient's suffering, the suffering is presented as bad for the patient. Second, even when the language used is ostensibly value-neutral, physicians may convey in a host of other ways, such as body language, tone of voice, and emphasis in presenting information, their own positive or negative evaluation of various alternatives. Third, physicians nearly always couple the process of informing the patient about treatment alternatives with a recommendation that a particular alternative be pursued. Moreover, providing a recommendation is not simply an unfortunate slip from the role of value-neutral information provider, but is understood by both physician and patient as a proper and important part of the physician's professional role and responsibility. For a variety of well-known reasons, sick patients are often extremely vulnerable to being strongly influenced by these treatment recommendations. Fourth, physicians commonly see a part of their role as securing patient compliance, which often requires them to be strong advocates for the selected treatment. Fifth, since it is never possible in practice to provide all possibly relevant facts about treatment alternatives, a value judgment must always be made about which facts are most important.

The respects in which physicians commonly fail in practice to conform to a value-neutral role in professional relations with patients are probably of greatest importance in evaluating the actual relations and the roles of facts and values in them. However, the facts/values division of labor might still be defended by maintaining that at least *in principle* physicians can and should maintain the role of value-neutral facts provider, even if in practice they often fail to do so. The very possibility and/or desirability, even in principle, of the value neutrality of physicians can be challenged at several levels.

The first level is in the most basic scientific theories of physical chemistry and biological science on which medical science builds. Is theory choice at this level fully determined by the facts about the world, as a simple empiricist view would have it, or is theory choice not fully determined by the world and so inevitably influenced by values, whether they be scientific values or broader social values? Are the concepts employed in these scientific theories value-free or do they carry value commitments? Is the truth of major alternative scientific theories and paradigms, and changes in them over time, dictated solely by facts about the world or are such alternatives instead incommensurable and dictated in part by values and other noncognitive factors?

These questions raise extremely complex scientific, philosophical, and historical issues in the philosophy of science. There is no consensus on what is the correct view, although few defend a simplistic empiricism according to which science and scientific theory are fully governed by a value-free and concept-neutral world and data. But this is not the level at which the physician's role most clearly embodies value commitments and so I will set these basic issues in the philosophy of science aside. Even if basic physical chemistry and biological science are in some significant sense value-neutral and value-free, that still leaves us a far distance from the conclusion that the physician's role in shared decision making is, or could be, value-neutral.

Let me make a relatively crude distinction between the basic physical and biological sciences on which medicine builds and medical science itself. It is fairly widely agreed that medical science is concerned, in the most general terms, with health and disease, or the normal and pathological, in human beings. There has been considerable debate in the philosophy of medicine about whether the concepts of "health" and "disease" are value-free empirical or factual concepts, or whether instead the determination of the healthy and diseased inevitably involves appeal to certain values or value judgments.

Defenders of a naturalistic, value-free theory of health and disease like Christopher Boorse and Leon Kass generally appeal to an understanding of health and disease in terms of normal and abnormal function of the organism.[8] Taking disease in an organism to be roughly equivalent to the pathological, Boorse (1987, p. 370) argues:

> the essence of the pathological is *statistically species-subnormal part-function*, or more carefully: A condition of a part or process in an organism is *pathological* when the ability of the part or

72

process to perform one or more of its species-typical biological functions falls below some central range of statistical distribution for that ability in corresponding parts or processes in members of an appropriate reference class of the species.

Functions of an organ, in turn, can be understood as "its species-typical causal contributions to the organism's survival and reproduction" (Boorse 1987, p. 371).

Critics of this view hold that the classification of a particular condition as healthy or diseased is not determined solely by functional, biological facts in nature, but requires imposing a value judgment on facts about conditions and functions.[9] In this view, diseases are not simply deviations from species-typical functions, but are deviations that are judged *bad* for the organism. These critics point to admittedly extreme historical examples in which masturbation, the desire of slaves to run away from their owners, and homosexuality have been classified as diseases on the basis of an evaluation of these practices, desires, and conditions as bad for their possessors (Engelhardt 1974).

In this view, the classification of homosexuality as a disease reflected a negative social evaluation of the condition. The American Psychiatric Association eventually removed homosexuality from the classification of disease because of ethical and political challenges, not because new scientific facts had been discovered.[10] Even in more common everyday cases of disease, it is argued, mere deviation from normal function does not entail disease when the deviation is in a positive direction or has a positive effect on the organism's function. Moreover, normal function is not just statistically normal function, but valued function, and concerns more than just survival and reproduction.

If determination of what constitutes well-functioning requires, at least implicitly, appeal to value judgments, and may vary in different social contexts and historical periods, then "health" and "disease" seem not to be value-free concepts. The import of this debate for the understanding of the physician's role in the division of labor of shared decision making should be evident. If the physician's role is to determine and communicate to the patient whether the patient is diseased, and what treatments of that disease are possible, the physician could not do this in a value-free way if the definition of what constitutes a disease is not a value-free scientific determination. If, instead, a particular condition is classified as a disease in part because it is considered bad for those who have it, then provision of information to a patient about his or her diseased condition implicitly invokes and carries with it the value judgment about its badness.

Whether or not the concepts of "health" and "disease" are themselves value-neutral, physicians commonly do understand themselves, in their role as medical professionals, as committed to preserving, promoting, and restoring health, and preventing and treating disease. Because the professional commitment of physicians includes a central commitment to the value of health and the disvalue of disease (even if it is possible to understand "health" and "disease" themselves in value-neutral terms), physicians cannot be understood simply as value-neutral providers of the facts. Their provision of the facts must be understood as being in the service of the value of their patients' health. Moreover, this is not simply an unfortunate deviation by physicians from the ideal of value neutrality; the role of advocate for their patients' health is a central part of common ideals for physician-patient relations and treatment decision making.

More generally, the professional understanding of the role of physician is substantially defined by ethical commitments which are, to a significant degree, special to medicine and not found in other social and commercial relations. Unlike commercial relations in which profit maximization is accepted as a proper motivation, the professional ethos of physicians requires that the interests of their patients must first and foremost guide physicians' behavior. This altruistic commitment of physicians to their patients' interests, and the concomitant setting aside of the interests of all others, including their own, is perhaps the most striking defining norm of medicine. The requirement of gaining the patient's informed consent for treatment likewise adds to the commitment to the patient's health and well-being, the ethical commitment to respect the patient's self-determination. The requirement to be truthful with and not to lie to patients about their medical conditions and treatment is a further ethical commitment that sets the medical profession apart from most other social and commercial relations.

The point is not that we don't expect any truthfulness elsewhere, but that the strength of the physician's ethical commitment to truthfulness with his or her patients is almost without parallel. Related to truthfulness is a very strong requirement of confidentiality governing information provided by the patient to the physician; this too must be seen as an ethical commitment contributing to the defining norms of what it is to be a physician. Finally, there are professional norms of medicine governing access to health care. Access to at least basic health care should not be determined by ability to pay, but should be determined by medical need.

Of course, all of these ethical norms are far more complex and controversial than the simple formulation I have given here would indicate. It should also go without saying that not all physicians always live up to and act in accordance with all of these professional norms. The point is rather that physicians commonly understand that being a physician entails a commitment to a related and coherent body of ethical norms by which the profession in part judges itself and asks others to judge it. These norms give substance to the general ethos of the medical profession as a "high calling." An ethically driven calling must be understood as serving an important good, and not entirely ethically neutral about the results of the practice of the profession. That important good is health, seen as ultimately in the service of the patient's well-being.

Now it might be argued that this ethically defined role of the physician still requires physicians to attempt as much as possible to be value-neutral presenters of the facts to patients in treatment decision making. The ethical definition of the role of physician might itself require striving for value neutrality. But I think that cannot be correct. The various ethical commitments that I have suggested substantially define the physician's role do not make sense except as in the service of important values. The central value, I have suggested, is health, seen as ultimately in the service of the patient's well-being.

To say that the value is "health, ultimately in the service of the patient's well-being," is to underline that the medical intervention best for a particular patient need not always be what is best for other patients with similar medical conditions. As noted earlier, different human functionings that are a part of the good for persons fall at different points on an objective–subjective continuum.

The more subjective the value of the functioning affected, the more advocating for patients' health is perfectly compatible with, indeed requires, a recognition that treatment that best serves "health, ultimately in the service of the patient's well-being," may differ for different patients suffering from the same disease.

I believe physicians are correct in seeing themselves as advocates for their patients' health. As patients, or potential patients, we want them to advocate for our health, especially in light of the degree to which serious illness can sometimes impair our own ability to be advocates for our health. This advocacy, however, must be seen as limited by respect for the self-determination of competent patients. It is this interest of patients in making important decisions about their own lives that requires physicians to respect and not to interfere with patients' treatment choices, even if those choices will be bad for them. If a patient's decision-making capacities are sufficiently defective to warrant a determination that he or she is incompetent to decide for himself or herself, a surrogate must make the decisions. The two values of patient well-being and self-determination together require a balancing of the physician's role between value-committed, non-value-neutral advocacy for the patient's health and a willingness to accept a choice that fails to best secure the patient's health.

76

Nevertheless, there will be some choices made by the patient that are in sufficiently serious conflict with the value of health to which the medical profession is committed that the physician may reasonably be unwilling to participate in carrying out that choice and must withdraw from the care of the patient. When patients withdraw from some or all of the care of their physicians, with likely serious adverse consequences for their health, those physicians are not, nor would we want them to be, merely value-neutral onlookers. Instead, they are active advocates with those patients for the patients' health, and when this breakdown in the physician-patient relationship occurs in the hospital, the patient must sign out "AMA" – against medical advice. This practice is not just for the legal protection of the physician and the hospital, but reflects the value-committed role as advocates for our health that we want our physicians to play.

Shared decision making does not imply a value-neutral role for physicians; it requires of them a more delicate balancing. They must act as advocate for their patients' health and well-being, while also being prepared ultimately to respect patients' self-determination, even when they disagree with their patients' treatment choices.

NOTES

1   See, for example, President's Commission for the Study of Ethical Problems in Medicine and Biomedical and Behavioral Research (1982).

2   One of the classic and most influential general accounts of positivist positions is Alfred J. Ayer, *Language, Truth, and Logic* (1936).

3   Perhaps the most influential account of ethics along these lines was Charles Stevenson, *Ethics and Language* (1944).

4   Though for an account of how far we remain from full patient participation in treatment decision making, see Jay Katz, *The Silent World of Doctor and Patient* (1984).

5   The difficulties in both intuitionism and naturalism will be familiar to philosophers. Those unfamiliar with them can find an account of these difficulties that places the development of

both intuitionism and naturalism in historical context in W. D.
Hudson, *Modern Moral Philosophy* (1970).

6 This is the characterization of subjectivism employed by Brink,
Moral Realism and the Foundations of Ethics (1989, pp. 217–36).

7 I discussed these theories some years ago as alternative accounts of utility in "Recent Work in Utilitarianism" (1973). A
more recent treatment is in my "Quality of Life Measures in
Health Care and Medical Ethics," in *The Quality of Life* (1992).
One of the best accounts of these alternative theories and the
difficulties of each is James Griffin, *Well-being* (1986).

8 Christopher Boorse's account is developed in: "On the Distinction between Disease and Illness" (1975); "Health as a Theoretical Concept" (1977); "Concepts of Health," in *Health Care
Ethics* (1987). Leon Kass's account is in his *Toward a More Natural Science* (1985).

9 For example, Peter Sedgwick, "Illness – Mental or Otherwise"
(1973).

10 See Ronald Bayer, *Homosexuality and American Psychiatry* (1981).

REFERENCES

Ayer, Alfred J. 1936. *Language, Truth, and Logic.* New York: Dover
Publications.

Bayer, Ronald. 1981. *Homosexuality and American Psychiatry.* New
York: Basic Books.

Bayles, Michael. 1978. Physicians as Body Mechanics. *Contemporary
Issues in Biomedical Ethics.* ed. John W. Davis, Barry Hoffmaster,
and Susan Shorten. Clifton, N.J.: Humana Press.

Boorse, Christopher. 1975. On the Distinction Between Disease and
Illness. *Philosophy and Public Affairs.* 5:49–68.

———. 1977. Health as a Theoretical Concept. *Philosophy of Science.*
44:542–71.

———. 1987. Concepts of Health. *Health Care Ethics.* ed. Donald
VanDeVeer and Tom Regan. Philadelphia: Temple University
Press.

Brink, David. 1989. *Moral Realism and the Foundations of Ethics.* Cambridge: Cambridge University Press.

Brock, Dan W. 1973. Recent Work in Utilitarianism. *American Philosophical Quarterly.* 10:241–76.

———. 1992. Quality of Life Measures in Health Care and Medical

Ethics. *The Quality of Life.* ed. Amartya Sen and Martha Nussbaum. Oxford: Oxford University Press.

Buchanan, Allen B. 1991. The Physician's Knowledge and the Patient's Best Interest. *Ethics, Trust, and the Professions: Philosophical and Cultural Aspects.* ed. Edmund D. Pellegrino, Robert M. Veatch, and John Langan. Washington, D.C.: Georgetown University Press.

Daniels, Norman. 1979. Wide Reflective Equilibrium and Theory Acceptance in Ethics. *Journal of Philosophy.* 76:256–82.

Engelhardt, Jr., H. Tristram. 1974. The Disease of Masturbation: Values and the Concept of Disease. *Bulletin of the History of Medicine.* 48(2):234–48.

Goodin, Robert. 1986. Laundering Preferences. *Foundations of Social Choice Theory.* ed. Jon Elster and Aamand Hylland. Cambridge: Cambridge University Press.

Griffin, James. 1986. *Well-being.* Oxford: Oxford University Press.

Hudson, W. D. 1970. *Modern Moral Philosophy.* Garden City, N.Y.: Doubleday.

Kass, Leon. 1985. *Toward a More Natural Science.* New York: Free Press.

Katz, Jay. 1984. *The Silent World of Doctor and Patient.* New York: Free Press.

President's Commission for the Study of Ethical Problems in Medicine and Biomedical and Behavioral Research. 1982. *Making Health Care Decisions.* Washington, D.C.: U.S. Government Printing Office.

Rawls, John. 1971. *A Theory of Justice.* Cambridge, Mass.: Harvard University Press.

Sayre-McCord, Geoffrey. 1988. *Essays on Moral Realism.* Ithaca, N.Y.: Cornell University Press.

Sedgwick, Peter. 1973. Illness – Mental or Otherwise. *Hastings Center Studies.* 1:30–31.

Sherlock, Richard. 1986. Reasonable Men and Sick Human Beings. *American Journal of Medicine.* 80:2–4.

Stevenson, Charles. 1944. *Ethics and Language.* New Haven: Yale University Press.

Veatch, Robert M. 1972. Models for Ethical Medicine in a Revolutionary Age. *Hastings Center Report.* 2(June):5–7.

Chapter 3

When competent patients make
irrational choices
Coauthored by Steven A. Wartman

In recent years, physicians and patients have tended to move toward shared decision making. Although it sounds reasonable on the surface that patients and physicians should collaborate in making decisions about medical care, surprisingly little attention has been given to the complex and troubling issues that can arise. In particular, what does shared decision making imply for a physician's responsibilities when an apparently competent patient's choice appears to be irrational? A discussion of this issue requires a taxonomy of the different sources and forms of irrational decision making. We believe such a taxonomy should include the bias toward the present and near future, the belief that "it won't happen to me," the fear of pain or the medical experience, patients' values or wants that make no sense, framing effects, and conflicts between individual and social rationality. Our main aim here is to develop this taxonomy and thus to bring out some of the theoretical and practical obstacles involved in distinguishing between a patient's irrational choices, which the physician may seek to change, and merely unusual choices that should be respected. To avoid any misunderstanding, we emphasize at the outset that even the irrational choices of a competent patient must be respected if the patient cannot be persuaded to change them.

SHARED DECISION MAKING BETWEEN
PHYSICIAN AND PATIENT

Historically, the professional ideal of the physician–patient relationship held that the physician directed care and made decisions about treatment; the patient's principal role was to comply with "doctor's orders." Although this paternalistic approach often took account of the patient's general preferences and attitudes toward treatment, it gave the patient only a minimal role in making decisions. When faced with what appeared to be a patient's irrational choices or preferences, physicians were encouraged by this approach to overlook or override them as not being in the patient's true interests.

Challenged by a number of forces within and outside the medical profession during the past two or three decades, the paternalistic approach has generally been replaced by the concept of shared decision making, in which both physicians and patients make active and essential contributions.[1] Physicians bring their medical training, knowledge, and expertise – including an understanding of the available treatment alternatives – to the diagnosis and management of patients' conditions. Patients bring a knowledge of their own subjective aims and values, through which the risks and benefits of various treatment options can be evaluated. With this approach, selecting the best treatment for a particular patient requires the contributions of both parties.[2]

This description of the division of labor oversimplifies the complexities of the roles and contributions of physicians and patients when real decisions about treatment are made, but it does highlight the patient's new, active part in that process. Some have concluded that in shared decision making, proper respect for patient autonomy and self-determination means accepting the patient's treatment preferences however they are arrived at. We believe that such a conclusion is unwarranted, because it fails to recognize the trade-off between the sometimes conflicting values that underlie shared decision making and that are involved in respecting or seeking to change patients' choices. The first value is the well-being of patients, which can require the physician to attempt to protect them from the harmful consequences of their choices when their judgment is irrational. The second value is respecting the right of patients to make decisions about their own lives when they are able. Whenever competent patients appear to be making irrational choices about treatment that are contrary to their own well-being, the two values will be in conflict.

Distinguishing choices that are truly irrational from those that are merely unusual often requires complex, difficult, and controversial judgments. When the physician properly judges a patient's treatment choice to be irrational, attempts to change that choice through persuasion are common and proper. Noncoercive and nonmanipulative attempts to persuade patients of the irrational and harmful nature of their choices do not violate their right of self-determination. Instead, they reflect an appropriate responsibility and concern for the patients' well-being.

Sometimes, however, attempts to persuade will fail. Physicians lack both ethical and legal authority to override patients' treatment choices unilaterally. Nevertheless, in a few cases an irrational choice that cannot be changed by persuasion may reflect a sufficiently serious impairment in decision making – and the consequences of that choice may be sufficiently harmful to the patient – to call the patient's competence into question. In such cases, the physician may begin an investigation of the patient's competence that can ultimately involve recourse to the courts. Since the vast majority of irrational decisions are made by apparently competent patients, we shall not address the determination of incompetence here.[3] Our concern is with the more usual cases, in which the patient's competence does not come into question. The responsibility to try to change their competent patients' irrational choices requires that physicians gain a better understanding of the different forms of irrational treatment choices and of the theoretical and practical difficulties involved in distinguishing truly irrational from merely unusual choices.

THE STANDARD OF RATIONAL DECISION MAKING

Any discussion of irrational decision making must rely on a description of rational decision making.[4] We believe it will be helpful to make that description explicit, if only in brief outline. Specifically, what is the norm of rationality that underlies the ideal of shared decision making between patients and physicians? Essentially, shared decision making entitles patients (or their surrogates if they are incompetent) to weigh the benefits and risks of alternative treatments, including the alternative of no treatment, according to their own values and to select the alternative that best promotes those values. In the language of decision theory, each patient's values will determine his or her utility function, and the rational choice will be the one that maximizes expected utility. Since treatment decisions always involve some degree of uncertainty about the beneficial and harmful effects of alternative treatments, these effects should be discounted by their probabilities (to the extent that they are known) in calculating the expected utility of various treatment alternatives. If the probabilities are not known, each patient's attitude toward risk will determine the weight given to uncertain beneficial or adverse effects.

Shared decision making requires that physicians ensure that their patients are well informed.[5] Thus, another aspect of rational decision making is that each patient has and uses correct information about relevant alternatives. This sketch of rational decision making relies ultimately on the patient's own aims and values, unless they are irrationally distorted in the ways discussed below, as the ends that guide decision making. An irrational choice is one that satisfies those aims and values less completely than another available choice.

There is a second notion of irrational decision making that deems a patient's choice irrational if it fails to promote a set of basic aims and values that belong to the physician or to standard guidelines of medical practice. Physicians who criticize a patient's choice as irrational in this sense are disagreeing with the basic aims and values by which the patient defines his or her own good, rather than arguing that the patient's choice will fail to promote those aims and values best. Since this second notion ignores the patient's own aims and values and thus fails to respect the right of self-determination, we rely here on the first account of rational and irrational choices.

FORMS OF IRRATIONAL DECISION MAKING

We now turn to common forms of irrational decision making by patients or their surrogates (and sometimes by physicians). In many treatment decisions, more than one form of irrationality affects a single choice, but we separate them here for analytical clarity.

Bias toward the present and near future

The ideal of rational decision making gives equal weight to a beneficial or harmful effect whenever it occurs in a person's life, with differences determined only by the size and probability of the effect. In the case of money it is rational to apply a discount rate, because a dollar received today can earn interest and is thus worth more than a dollar received 10 years from now. Some effects of health care are similar: it is rational to prefer a restoration of function now rather than later and to prefer that a loss of function occur as far in the future as possible, so as to minimize the period of disability. Similarly, it is rational to prefer that the loss of one's life be postponed for as long as possible, at least while it remains a life worth living. For other effects of medical care – especially pain and suffering – rational choice would seem to require indifference to their timing. In particular, it is irrational to refuse to undergo a painful experience now, if by undergoing it one can avoid a much worse experience in the future. Such a refusal would amount to preferring more rather than less pain or suffering in one's life.

Yet, as doctors know, medical practice is replete with such irrational choices by patients. Some patients who continue to smoke or drink heavily or who fail to follow relatively simple steps to control moderate hypertension may not be irrational, but are simply willing to gamble that they will beat the odds. Others, however, have given inadequate weight in their present decision making to the harm they are likely to suffer in the relatively distant future. We call this a bias to the present and near future, because people commonly give disproportionate weight to securing benefits and avoiding harm in the present and near future as opposed to the more distant future.[6] The physician's task in such cases is to help the patient fully appreciate the size and seriousness of the more distant harm or benefit, so that it can play an appropriate part in the patient's decision making.

"It won't happen to me"

Patients may view the nature of the risk or harm of not following medical advice differently. This is especially true for events that have a low probability of occurring.[7] However, what constitutes low probability may vary considerably from patient to patient. Furthermore, since some patients are more willing to take risks than others, it is often difficult to determine whether a patient is more of a risk taker than most or whether the patient has simply failed to give adequate weight to a low probability or a distant event. This situation is complicated by the difficulty of distinguishing among patients who, for example, ignore a risk (that is, acknowledge the risk but decide to accept it), irrationally deny the possibility that an untoward event could happen to them, have "magical" or illusory beliefs about their vulnerability to harm, or simply have a different way of viewing the medical problem.[8] Adolescents, for example, are commonly subject to feelings of invulnerability to certain harms disproportionate to the real risk of those harms.

Physicians often need to gain some understanding of their patients' general attitudes toward risk and the extent to which they are risk-averse, perhaps as evidenced by their past behavior. Physicians should attempt to distinguish among the possibilities noted above. Sometimes a physician can help a patient appreciate a risk more vividly and relate it to the patient's life. However, for patients who deny a risk or have magical beliefs, a more detailed medical and scientific explanation is not likely to be helpful. In these cases, formal counseling or psychiatric evaluation may be more fruitful.

Fear of pain or the medical experience

Many patients delay or will not even consider a particular treatment for fear of the perceived nature of the experience, although they may acknowledge that the treatment is clearly in their best interests. Sometimes their decision is coupled with some form of rationalization – "there's no need to do it yet," or "I'm too busy now with other things," for example. In other cases, when a dreaded experience draws near, a patient may be almost paralyzed with fear. Sometimes the fear may be focused not on pain or suffering but on other dreaded experiences, such as being "cut open" or being "put to sleep" in surgery. In still other cases, the fear of a disease such as cancer or the acquired immunodeficiency syndrome can prevent a person from making informed decisions about its treatment.

Determining when this form of irrational decision making is present is considerably complicated by the fact that no single, correct weight can be given to pain or a particular medical experience as measured against the beneficial outcomes for which the experience may be necessary. Patients differ, for example, in the degree to which they are prepared to tolerate painful treatments or conditions for the sake of other ends,[9] but these reasonable differences are difficult to distinguish from the undue weight some patients give to certain aspects of treatment because of irrational fear. Physicians may have seen patients who in the end were grateful that they had been pressured or even forced to undergo painful or dreaded treatments. The responsibility of the physician in these cases is a difficult one – to respect the different weights people give to avoiding pain and suffering, while helping patients overcome the irrational fear that prevents them from pursuing promising treatment plans. The physician's task will often involve helping patients to distinguish whether they are experiencing a fear that they want to overcome or whether they have made a choice with which they are comfortable.

What the patient wants does not make sense

When competent patients decline a recommended course of treatment because of an obvious and understandable, albeit unusual, belief – Jehovah's Witnesses, for example, who refuse blood transfusions – physicians (and the courts) commonly yield to that belief. When patients request treatment that physicians believe to be ineffective – Asian patients who request "coining," for example – physicians are not obliged to provide it, but they may respect the patients' right to pursue it when medically acceptable treatment is also provided. Special difficulties arise when a seemingly competent patient wants something that does not make sense and is not attributable to a clearly recognizable religious belief or cultural preference. It can be extremely difficult in these situations for the physician to determine the basis of the patient's preference. However unusual, the more the preference seems to reflect a deeply held, enduring value that is important in the patient's life, the stronger the case for respecting it, as long as it does not require that the physician participate in useless or medically unacceptable treatment.

In other cases, what the patient does not care about makes no sense. For example, patients may state that they understand but simply do not care that death or serious disability will result from a refusal of treatment. It may be difficult to determine whether this is an authentic, although unusual, choice, or the result of a distortion of values caused by a treatable condition such as depression.

Framing effects

It is well known that the way choices are formulated and presented, or framed, can have major effects on decisions.[10] A simple example is the presentation of a surgical treatment as "substantially extending the lives of 70 percent of the patients who select it" or as "potentially killing on the operating table 30 percent of the patients who select it." Both characterizations may be true, but which is used, or emphasized, may have a substantial effect on the rate of selection of the surgery. There is a variety of different and more subtle framing effects, one of which we illustrate below.

Studies in the psychology of choice show that losses tend to loom larger than gains in most people's decision making. Of course, whether a particular outcome is viewed as a gain or loss depends on the reference point against which the outcome is compared. Many choices in medicine can be framed in either way. For example, lowering moderate hypertension can be presented as adding months to the patient's expected life span or as a way to avoid shortening the life span because of untreated hypertension. Neither framing of the choice before the patient is wrong; each simply relies on different characterizations of the patient's present situation. Tversky and Kahneman[11] have compared the framing effects in decision making to changes in perspective in visual judgments. Which of two mountains appears to be higher, for example, depends on the position from which one views them. There is, of course, an objective standard by which the height of the mountains can be determined, but there appears to be no objectively correct way to frame many medical choices, such as that facing the patient with moderate hypertension. There are simply two different but correct ways to frame the choice, and the one that is used will influence whether some patients choose treatment. Sometimes, the best that physicians can do is present the choice in alternative ways in the hope of minimizing framing effects.

Individual versus social rationality

The irrational use of resources. The circumstances that make individual choices rational can sometimes make the outcome of those choices irrational when viewed from a different perspective.[12] One factor fueling the intense pressure to control rapidly rising health care costs is the perception that resources are often used in circumstances in which their expected benefits do not justify their true costs. An insured patient has little or no economic incentive to weigh the true costs of medical care against its expected benefits. When patients have no out-of-pocket costs, it is rational for them to choose all the care that has any expected medical benefit, no matter how small or costly. If physicians accept the common professional norm that their obligation is to do whatever may benefit their patients, without regard to cost, then it is also rational for them to ignore the cost of care in making recommendations and decisions about treatment. The result will be the overuse of health care as compared with other goods and services whose benefits are weighed against their true costs. From the perspective of those who pay the insurance premiums (employers or the government, for example), the result is an irrational overallocation of resources to health care.

The issues raised by this form of irrational social choice involving the use of resources are very different from those involved in the forms of irrational patient choice previously discussed. It would be a mistake for physicians to seek to persuade insured patients that choosing care that is not cost-effective is irrational. On the contrary, an insured patient's choice of such care is rational, but it leads to an irrational overallocation of resources to health care. Since the irrationality is not at the level of the insured patient's choice of treatment, the principal response to it cannot be at that level.

A physician's failure to respect an insured patient's choice of such care on the grounds that it is irrational is not justifiable. Instead, the irrationality must be addressed where it exists – in the social and economic system of health care financing.

Public health versus individual benefit. Often, physicians are concerned about the public health benefits of medical interventions, whereas their patients are not. For example, national campaigns to reduce serum cholesterol levels will clearly benefit the health of the country as a whole. However, the chance of a substantial benefit in a given patient may be very small. Consequently, some people may rationally decide that for them the benefits of the intervention do not outweigh its burdens. This distinction between community-wide and individual benefits has been called the "prevention paradox"; in it, a treatment that brings large benefits to the community may not seem worth the trouble to individual participants.[13] There is no true paradox, however. The existence of a society-wide benefit constitutes no reason to view as irrational an individual's choice to decline an intervention.

For some infectious diseases, preventing infection through the vaccination of one person (or shortening the period of transmissibility through treatment) lessens the risk of disease for others. A patient's or parent's refusal to accept immunization may be rational if the patient or parent is not concerned about the risks for others or believes that because enough of the population is immunized, the threat of the disease is minimal and the risks of immunization outweigh the benefits. In this case, for the protection of others, society may adopt mandatory immunization programs or physicians may seek to change the patient's mind. Patients do not have an unqualified right to make even rational individual choices that risk serious harm to others.

WHAT SHOULD PHYSICIANS DO?

Shared decision making respects the patient's right of self-determination but does not require that the patient's preferences be simply accepted when they are irrational. In most cases, it is appropriate for physicians to attempt to persuade competent patients to reconsider their irrational choices. However, distinguishing irrational preferences from those that simply express different attitudes, values, and beliefs can be difficult in both theory and practice. Physicians need to be sensitive to the complexity of these judgments in helping patients to make sound treatment choices. They must also bear in mind that even truly irrational choices are not sufficient to establish a patient's incompetence and to justify overriding them. The taxonomy of irrational treatment choices we have presented here (and expand on elsewhere[14]) is meant to be a beginning guide for further consideration of the issue. More research is needed on the frequency of irrational treatment choices and their different forms, as well as on how physicians and patients can work together to overcome them.

NOTES

1　President's Commission for the Study of Ethical Problems in Medicine and Biomedical and Behavioral Research. Making health care decisions: the ethical and legal implications of informed consent in the patient-practitioner relationship. Vol. 1. Washington, D.C.: Government Printing Office, 1982.

2　Forrow L, Wartman SA, Brock DW. Science, ethics, and the making of clinical decisions. JAMA 1988; 259:3161–7.

3　Buchanan AE, Brock DW. Deciding for others: the ethics of surrogate decision making. Cambridge: Cambridge University Press, 1989.

4　Pauker SG, Kassirer JP. Decision analysis. N Engl J Med 1987; 316:250–8.

5　Katz J. Why doctors don't disclose uncertainty. Hastings Cent Rep 1984; 14(1):35–44.

6　Parfit D. Reasons and persons. Oxford: Oxford University Press, 1984.

7　Tversky A, Kahneman D. Judgment under uncertainty: heuristics and biases. Science 1974; 185:1124–31.

8　Gillick MR. Talking with patients about risk. J Gen Intern Med 1988; 3:166–70.

9   Cassell EJ. The relief of suffering. Arch Intern Med 1983; 143:522–3.
10 Tversky A, Kahneman D. The framing of decisions and the psychology of choice. Science 1981; 211:453–8.
11 Ibid.
12 Menzel PT. Medical costs, moral choices: a philosophy of health care economics in America. New Haven, Conn.: Yale University Press, 1983.
13 Rose G. Strategy of prevention: lessons from cardiovascular disease. BMJ 1981; 282:1847–51.
14 Kassirer JP, ed. Current therapy in internal medicine. 3rd ed. Philadelphia: B.C. Decker, 1990.

Part II

Life-and-death decisions in the clinic

Chapter 4

Moral rights and
permissible killing

In what types of cases that arise in the practice of medicine is it morally permissible to kill human beings? My discussion of this question will largely involve working out some implications of a particular type of normative moral theory for the question of terminating human life in medical settings. However, the very substantial and often heated disagreement that arises over these issues amounts to more than simply disagreement over the implications of the kind of view I shall develop. Rather, it arises in large part because participants in the disputes hold general moral views of fundamentally different sorts. In the emotionally charged context in which these discussions often take place, and where the specifics of particular cases often occupy much of the attention, the deeper sources of the moral disagreement encountered are often obscured. I shall begin then with a brief attempt to illuminate some of those sources.

I

How one divides and distinguishes substantive moral theories is largely a matter of convenience, in the sense that it depends on the purpose of the inquiry at hand and the usefulness to that inquiry of carving up the geography of normative ethics in that particular way. No general distinctions between types of normative theories which delineate broad categories of such views are *the* correct distinctions or categories; any quite general distinctions will at the very least

95

blur differences that are important in other contexts. Moreover, though I shall talk of three different types of moral theories with regard to their implications concerning the killing of human beings, it is not the case that we can neatly divide up real persons into holders of one or another of these types of moral view. Rather, for many if not most persons, two or even all three of the different types of moral positions have some appeal and place in their moral reasoning; this helps account for the uncertainty and internal conflict many of us experience in considering concrete cases in this area.

The first sort of moral view is what Ronald Dworkin has called goal-based and what is more commonly termed consequentialist, utilitarian, or teleological.[1] Most generally, it views morality as concerned with the production of desirable or valuable states of affairs or experiences – commonly human happiness, welfare, or desire satisfaction; their production is the goal of morality. Human actions are morally evaluated in terms of their tendency to promote these goals, and right action is that action which, among the alternatives open to an agent, maximizes these valuable consequences for any and all persons affected. The consequences for any one person count for no more than for any other simply because of who that person is; it is a matter of the amount of, for example, human happiness or desire satisfaction produced, and not of who produces it or who receives it. The morally significant features of persons in the goal-based moral view are simply their roles as subjects, recipients, or vessels of the experiences that the theory holds to be valuable; so long as the causal consequences for morally valuable experiences remain the same, persons from this moral point of view are fully interchangeable in any situation. In this goal-based view, killing a human being is morally justified if and only if doing so maximizes the production of the goals of the theory, however they are specified, and is morally wrong if it does not; killing is morally evaluated according to its production of the goals the theory specifies as valuable.

The two other sorts of moral theory are what philosophers

commonly call deontological. Each makes an appeal at the most basic level to a different sort of moral notion, whose logical form carries implications for the sort of actions that are morally justified according to the theory as well as for the general focus the view employs. These are what Dworkin has labeled duty-based and rights-based moral views. Both provide a focus, though of different sorts, directly on human deliberation, choice, and action which is not derived solely from the consequences of that action for the production of valuable experiences. A duty-based moral view posits a moral ideal for a person in the form of a set of moral commitments, constraints, or prohibitions of behavior that can be violated only at the cost of one's moral integrity, at the cost of becoming morally evil or corrupt. This kind of moral position focuses directly on individual decision and action, and evaluates that decision and action not solely on the consequences it brings about. It might be thought of as positing certain limits on how one can act toward others beyond which a person cannot pass without becoming evil. Such prohibitions will usually concern what one deliberately does to other persons, what are sometimes loosely called one's projects, as opposed, for example, to what one simply fails to prevent. The prohibitions may be, though they need not be, absolute prohibitions which no conflicting moral consideration can justify violating.[2] The Catholic position that the direct killing of an innocent human being is always wrong is an example from moral theology, and the common view that deliberate torture of human beings is always wrong, is unthinkable for a moral person, serves as a secular example. The moral categories of a duty-based view are all constraints on our action to which we must conform at the peril of our moral integrity. These constraints commonly derive from the general idea of relations between persons, and the moral requirement that what we do to others be justified by features of that specific other, and our relation to him.[3] Duty-based views attempt to account for the intuition that a person has a special responsibility for what *he* deliberately and directly does to others, different from his responsibility for the actions of others. In this sense, duty-based views take the distinction between persons more seriously than do goal-based views, and focus on persons as agents whose own deliberate actions put them into morally significant relations with specific other persons. In most duty-based views, not killing human beings is a basic moral duty in the theory.

Finally, compare a view that appeals generally, and in particular with regard to the issue of killing, to moral rights that persons are asserted to possess. Rights function differently than duties in that they delineate areas in which the person possessing the right is at liberty to act as he sees fit and to act in his own interest as he understands it, as opposed to delineating specific constraints to which he must conform. Rights specify, as well, areas of behavior in which it is wrong, at least without special justification, for others to interfere with the rightholder's exercise of his right, or to fail to provide that to which the rightholder has a right if and when he calls for it. We may choose to exercise our rights or not as we see fit. Rights-based views emphasize a view of persons as capable of forming purposes, of making plans, of weighing alternatives according to how well they fulfill those plans and purposes, and of acting on the basis of this deliberation. Rights protect our exercise of these capacities, capacities whose exercise is often associated with the notion of autonomy, independent of how doing so promotes goals specified as valuable. We might say that while both duty- and rights-based views focus on individual action in a way that goal-based views do not, duties focus on constraints on action while rights focus on unconstrained choice. For a rights view, killing a human being will be wrong because it violates his right not to be killed; it wrongs the one killed, by violating *his* right, in a more direct manner than occurs on either a goal or duty view.

We might then summarize that in a goal-based view morality is concerned with the production of valuable experiences; in a duty-based view with the conformance of human action to prohibitions necessary to the maintenance of moral integrity; and in a rights-based view with the exercise and preservation of moral rights and the free choice they protect.

All of this is, of course, extremely sketchy and a great deal more could profitably be said in delineating and sharpening the nature of the three sorts of views and the many variants each may assume. I hope it is sufficient, though, to suggest the radically different implications for the issue of killing that the three positions have. All three views, we may assume, hold killing persons to be seriously wrong, but they differ on why it is wrong and in turn on when it is wrong. Consider the following cases:

Case A. Jones has a terminal, incurable disease. It completely prevents him from leading a normal life, causes him considerable though not unbearable pain and suffering, and he is expected to die from it within roughly a year; he has no relations or friends who care about him. Jones explicitly requests that everything be done to keep him alive as long as possible, even though his treatment is fantastically expensive and exhaustive of medical resources. Because of unique features of Jones' medical situation, if he is killed now it is likely that new medical knowledge will be obtained that will enable alleviation of the suffering of other similar patients.

Case B. Smith has terminal, incurable cancer. It completely prevents him from leading a normal life, causes him considerable though not unbearable pain and suffering, and he is expected to die from it in roughly a year. His treatment is expensive, but such that his family can afford it without undue stress. Smith is fully in command of his rational faculties, has given long and serious thought to his situation, and has decided he wants to die because life in his present condition is not worth living. He is unable, in his present situation, to bring about his own death, and requests another (his doctor, wife, etc.) to do so. He will only die if steps directly intended to kill him are taken.

If we take a goal-based view and act to maximize the balance of good consequences for all affected, then it seems reasonable to conclude that in Case A Jones ought to be killed. There seems no reason to suppose that cases of this sort, where the happiness of all affected will be increased by someone's being killed, do not occur with some frequency. Duty- and rights-based views will both hold killing Jones to be wrong in A, the former because doing so would violate the doctor's duty not to kill and so violate his moral integrity, the latter because Jones has a moral right not to be killed that he has neither waived, forfeited, or failed to exercise. Case B makes clear the difference in the rights- and duty-based views. For the rights view, Smith has waived his right not to be killed by explicitly requesting his doctor to kill him, and so it is morally permissible for his doctor to do so; no right of Smith's is violated and no wrong done him. For the duty view, the moral constraint is on how the physician (or wife) can permissibly behave to avoid moral corruption and avoid doing evil; that constraint need not be affected by the wishes of others. Smith would be viewed as asking the physician to act evilly, a request the physician is morally required to refuse, and which in no way need affect his duties in the case.

Much more could be said about these cases, filling in details and answering objections, to support the interpretations I have given them in the three different sorts of moral views, but I shall not attempt that here. I have only wanted to suggest how the *form* of the overall moral theory or view that one holds can lead people in substantially different directions on questions of killing within medical settings. As already noted, few of us fall neatly and completely into one of these three moral camps, but rather what is more often the case is that each of us feels at least some of the plausibility of each of these three ways of looking at the morality of killing; we experience what might be called *intra*personal disagreement on these issues, as well as *inter*personal disagreement. Most people's actual moral views are a complex of these three different sorts, the parts of which are in continual and often shifting tension. This intrapersonal complexity can come at the price of internal conflict and lack of coherence in one's overall moral views, which in turn can generate motivation for bringing all one's particular moral positions into conformity with one or another of the sorts of moral views I have delineated, as one way of regaining coherence in one's moral views.

It is, of course, quite natural to want to move from the delineation of different kinds of moral theories and their implications in a particular area such as killing, to the task of determining which is the correct, true, or most justified theory. In what sense if any this can be done is, of course, perhaps the central issue of ethical theory, but an issue which would take us far afield if considered here. This much, however, is worth adding. If one holds, as I do, a view on the justification of normative ethical positions similar in major respects to John Rawls's view of justification and reflective equilibrium, with its implications for a coherence account of normative ethical theories, then the task at hand is to work out the implications of different moral theories on various moral issues in order to match them up with what Rawls calls our considered moral judgments on those issues.[4] I am more inclined than I believe Rawls is to expect that the reflective equilibrium of different persons will not converge on a single view, but rather will tend to divide at least along the lines of the three different forms of moral views I sketched above. If so, it is all the more important to be clear about precisely where and how they do diverge. In the remainder of this essay I shall take up that task by exploring some implications of a rights-based view, a view that holds killing to be morally wrong because it violates a person's moral right not to be killed.

II

If we begin with a moral right of a person not to be killed, then the issue of morally permissible killings becomes a question of when that right is not sufficient for the moral wrongness of killing a human being. One class of case will be the human being's failure to have the properties of personhood necessary and sufficient for possessing the right not to be killed; he is not the sort of being who has the right at all. I shall not discuss this class of cases here. A second class of case is the person's having forfeited his right not to be killed. Killing in self-defense is the most prominent example, and in some views capital punishment falls into this class as well. However, I can think of few if any significant cases in the medical context that plausibly fit the forfeiting category, and so shall set it aside as well.[5] The third class of case is that in which the right-holder has waived his right not to be killed. This class is prominent in the medical context, and I shall begin with Michael Tooley's account of voluntary euthanasia in order to bring out what I believe is a more acceptable account of rights, the right to life, and so the category of voluntary euthanasia.[6] The fourth class of case is when a person's right to life, or right not to be killed, is justifiably overridden by some competing moral consideration. I shall argue that these cases require appeal to a theory of justice or fairness. Moreover, thinking of these cases in terms of rights justifiably overridden, as opposed to how conflicting interests are resolved or accommodated, may shed a rather different light on them. I should add before proceeding that, throughout the remainder of this chapter, I shall be considering cases in which terminating a person's life is *morally* justified, apart from either what the law at present allows, or ideally ought to allow. I believe that morality here should form the basis for legal policy, though it is, of course, not the case that what is morally permissible or prohibited should *ipso facto* be legally permissible or prohibited.

How should we understand the category of morally permissible termination of human life which Tooley labels voluntary euthanasia? He characterizes it as "cases in which a person exists, but his state is such that he has a rational desire that his life be terminated; that is, a desire to die that is in accordance with an estimate of his own long-term self-interest that is based upon the best information available."

The claim that the presence of a desire to die means that killing a person does not violate his right not to be killed is based on a general account of the nature of rights and of the conditions in which they are and are not violated. Tooley assumes, in my view correctly, that the right to life is like other rights, in its form or logic, though its content differs. But he has ignored two aspects of the nature of rights, and as a result has produced a defective account of voluntary euthanasia.

First, to avoid violating A's right to x when I deprive A of x, it is not enough that A simply not desire x. A must have consented or granted permission to me to deprive him of x; he must have waived his right to x. For example, suppose I have no desire to vote in the next election, and Jones removes my name from the list of eligible voters. Jones has wrongly deprived me of my right to vote even though I did not intend to exercise my right; I have a moral (and legal) complaint against Jones for violating my right, even though I had no desire to vote. He has deprived me of that to which I have a right without my authorization or consent – without my having waived my right. If this is correct, and if the right to life is understood as structurally similar to other rights, then to avoid violating my right to life it is necessary not just that I desire to die but that I have consented to be killed, or waived my right not to be killed.

By concentrating on the rightholder's present desire for the object of his right rather than whether or not the right has been waived, Tooley misses an important aspect of rights that adds to their usefulness and value in facilitating our control and prediction of our future. On his present desire analysis of the conditions under which rights are violated, if someone gives you permission to kill him in certain circumstances, you do not have a right to do so if, when he is actually in those circumstances, you find that he wants to go on living; it is not prior permission or consent to be killed, but the present desire to die, which would be morally necessary. I believe this is mistaken. The granting of permission or waiving of a right may, like the granting of authority to another to act for us or in our name, be either revocable or irrevocable; e.g., the law allows one to give up certain rights to property by establishing an irrevocable trust (where one's subsequent desire at future times for the trustee to exercise certain rights is not necessary to his having those rights) or a revocable trust (where such a continuing desire is necessary in the sense that without it, one may revoke the trustee's right to act with regard to the property). The waiving of one's right not to be killed in specified circumstances should like-wise be able to be revocable or irrevocable. A person might wish it to be irrevocable if, for example, he believed that human life should only be lived when one was in possession of certain capacities and faculties, and should not be pro-longed beyond that point, but believed that he might be, for various reasons, what could be called weak-willed in such circumstances, and would then have and act on a desire to continue living. In such a case, a person might have good reason to waive irrevocably his right not to be killed under specified circumstances. When those circumstances arose it would not be necessary for the desire to be present for it to be morally permissible for another to kill him. At least when our future decision-making capacities will be impaired, there is nothing morally objectionable, in itself, in our present self binding our future self in a way which will be contrary to the desire of the future self. This is a function in part of the fact that we each view ourself as one single self that continues over time, so that it is *my* future that my present action seeks to control.[7] Odysseus lashed himself to the mast for just such purposes, an alcoholic may keep no liquor in his house so that it is not available when he wants it in the future, and in fact persons often perform actions which are structurally of this sort. That particular rights may be waived irrevocably allows us to control our future in ways that would otherwise not be possible.

Many persons, of course, do not possess such an ideal, and realize that their desires may change with time and as conditions change. Thus, they have good reason to make their consent for another to kill them under specified circumstances revocable should they in fact not desire to die when in those circumstances, and especially since dying is the most irreversible of occurrences: one cannot, as with other events which may turn out to be unwanted, "make the best of it" and try to recoup one's loss, as with a bad bet or investment. With a revocable waiving of one's right not to be killed, then, the presence of the desire to die at the later time is relevant only in establishing that the person does not wish to revoke his waiving. As a practical matter, it seems that both irrevocable and revocable waivings of one's right not to be killed under specified circumstances are intended by different persons, for example in so-called Living Wills, and so it is important to be clear, as is usually not done, about which is intended. Some requests for euthanasia specify it only for conditions where one will then not be in a position to confirm or deny one's present desire to waive that right, and here the issue of revocability is not of practical importance.

A more important aspect of rights is that their possessor may waive them, or give consent for acts that would otherwise violate them, for reasons that need not accord with the best available estimate of his long-term interests. This issue is far too complex for a full treatment here, but let me at least roughly suggest how I believe the matter is correctly conceived. A person's interests are not simply equivalent to the objects of his present desires at any point in time; rather, he may be mistaken about his interests and so desire things not in his best interests. This is because while what a person desires is ascertained (in a rather complex way) by what he says he wants and what he pursues, what his interests are is ascertained (again, in complex ways) by what promotes his welfare or good, with the latter having an objective status independent to some degree of what he simply happens to desire. Consequently, other persons may at times be better judges of what best promotes our interests than we ourselves are. On the other hand, it is important that though the concept of interests has this objective status, what that objective content is, and

105

how it is determined, is often controversial, so that what best serves a person's interests is often a contested question. Tooley has suggested that a person's desire to waive his rights is rational only when that desire is in his overall best interest, and only when it is does it justify other persons acting in ways that would, in the absence of the desire, violate his rights. While he does not take up the issue, this account suggests as well that our exercise of our rights ought only to be protected from interference by others when that exercise is in our best interests. But if rights were understood in these ways, their character would be substantially different from what I believe it is, and something significant and valuable about rights would be lost. Rights protect our freedom to act in the areas the right covers even in ways that may not be in our overall best interests. This feature reflects first a view of persons as able to form purposes, weigh alternative actions according to how they fulfill those purposes, and act on the result of that deliberative process; and second, it reflects a value placed on our autonomy or self-determination in making such choices and directing our lives, a value that is not simply derived from our choices best promoting our welfare, happiness, or interests. That our view of rights in fact has this feature can perhaps best be shown with an example. Whatever the more basic moral right from which they are derived, Americans are considered to have a right to marry whom they wish and to choose the occupation they wish. Others may try to convince one that a particular choice of mate or job is a great mistake, that it is not in one's best interests; however, even if on the best evidence available they are clearly correct, one's right is still to make and act on one's own choice, to marry the wrong person and take the wrong job. It would be wrong for others to forcibly prevent one from doing so. These examples concern preventing a person from exercising his rights. Consider a case of waiving rights where it is not in the best interests of the rightholder to do so. My property rights make it the case that others would act wrongly in taking property that is rightfully mine, but if I choose to give large sums to a foolish cause even when it is clearly not in my best interests to do so, representatives of that cause do not violate my rights in taking it.

Now while rights protect our freedom to act in ways that need not be in our best interests, it is not the case that they are sufficiently strong to make interference in the exercise or waiving of a right taken in the interests of the rightholder never justified. The issue is, When does a rights-based view permit paternalistic intervention which infringes a person's rights? I can only sketch an answer here.[8]

In the areas of behavior which they cover, rights give protection to our uncoerced choice and action. As indicated above, an institution of rights presupposes that persons have capacities to form purposes, weigh alternatives, and act, and reflects in turn the value most of us place on being free to exercise these capacities without interference from others, a process we value for its own sake as part of an ideal of human excellence and not simply as instrumental to human happiness or some other goal. In this view, we will want others to act paternalistically towards us only when our *capacities* to form purposes, weigh alternatives, and act are seriously defective, only when we are clearly subject to a serious defect of reason or will. The forms such defects can take are various and I shall not say more about them here, except to emphasize that the mere judgment by another that one's express desires will not best serve one's interests, even where on the evidence that judgment may be considered by most to be correct, is not sufficient to demonstrate incompetence and to justify paternalistic intervention contrary to those desires. It is the freedom to decide for ourselves as we see fit what direction our life will take, even when others may with good reason disagree, that our moral rights are designed to protect.

Since paternalistic action must be taken for the good of the person involved, I am doubtful whether it is ever the case that killing a person could be justified on paternalistic grounds in the face of a strong and continuing desire on that person's part to live. The difficult issue is whether we should on paternalistic grounds discount his desire to die. Pressing this question further would require a full account of paternalism. But enough has been said to suggest why for a rights-based view a request to be killed need not be in the person's overall long-term interest to be a valid request, though there will be some cases where, because the person is not competent to exercise and waive his rights, paternalistic grounds may justify invalidating his request to die. The practical implication of this point is that in specifying when they wish euthanasia to be performed on them, persons should not be restricted by any particular conception of rational desires or of their long-term interests.

We can conclude that in standard cases of voluntary euthanasia, the desire to die must actually be expressed – that is, consent given to be killed so that one's right not to be killed is waived – but the desire or request need not always be a present desire and need not always be rational in the sense of best serving the person's long-term interest.

I want now to turn to cases where a person is unable to express his desire either to die or to go on living, unable to waive or insist on his right not to be killed. If we include as well cases where paternalistic intervention is justified because the person is no longer competent to exercise his rights, as in many cases of advanced dementia, then this class of case is probably of considerable practical importance. Is it reasonable to construe any such cases as instances of voluntary euthanasia, and if so which?

Tooley proposes one extension of the category of voluntary euthanasia. He concentrates on two cases of persons suffering from some extremely painful, incurable disease, a disease in which everyone who has it prefers death to life, and of whom it is known that they do not disapprove of euthanasia: a child who has not yet learned a language, and an adult who is paralyzed and cannot make his desires known to others. Both examples are designed to allow us to be virtually certain that the person in fact has, and will continue to have, a desire to die, but just cannot express that desire to us. On my account of the right not to be killed, and of the conditions under which killing does not violate that right, these are not cases of voluntary euthanasia because voluntary euthanasia requires consent to be killed, a waiving of the right not to be killed.

Involuntary euthanasia, on the other hand, is when a person is killed who is capable of waiving his right not to be killed should he so desire and explicitly declines or refuses to do so, or has not yet done so. These examples of Tooley's actually constitute a third sort of case – nonvoluntary euthanasia in which the potential victim is unable to give or withhold consent to be killed, unable to waive or decline to waive his right not to be killed. The plausibility of assimilating these examples to standard cases of voluntary euthanasia is that they suggest that we have sufficient evidence to be virtually certain that, were the person capable of waiving his right not to be killed, he would do so. I have no substantial quarrel with assimilating these cases to the category of voluntary euthanasia so long as it is clear that doing so is legitimate not just because of evidence of a desire to die, but because that evidence strongly supports the counterfactual that the person would waive his right not to be killed were it possible for him to do so.[9]

The more frequently encountered case, however, is a person who is capable neither of competently waiving or declining to waive his right not to be killed, where we have no decisive evidence regarding what he would do if he did not lack this capability. This case cannot be assimilated to the category of voluntary euthanasia, although it is more typical of what physicians and families actually encounter. It might well be wondered whether they ever encounter cases of the sort Tooley imagines, where virtual certainty exists about the person's desire though he is incapable of expressing the desire. Such certainty in his examples is based on the fact that everyone in similar circumstances who can express such a desire, in fact does so; but do we ever have evidence of this sort? It is well known that in many persons a desire to live may persist in the face of the most severe, incurable and incapacitating suffering; especially, but certainly not only, or even usually, with persons who believe euthanasia is wrong. Could we ever be certain that the case before us, where the person is incapable of consenting or declining to consent to be killed, was not of this sort? Surely, given the unpredictability of the strength of the will to live in any particular person, physicians must usually, if not always, be uncertain about whether a patient in cases of this sort would wish to and consent to be killed.

It is important to distinguish two sorts of cases in which a person is unable to make his wishes known – in the first he is undergoing extreme and continual suffering, whereas in the second he is relatively free from suffering. In the second case, and in the absence of a prior waiving of the right to life for situations of this sort, it is difficult to see what evidence could strongly support the belief that the right would be waived were it possible to do so. Since infringing upon such a fundamental moral right is an extremely serious moral wrong, I believe that in cases of the second sort one should assume that the right not to be killed would not be waived. Where the patient is undergoing extreme suffering as well, the case for a general policy of erring on the side of not violating the right to life, by assuming that the right would not be waived, is less persuasive. A moral theory asserting the existence of fundamental moral rights will quite likely include as well a right of persons to have extreme suffering relieved by others, at least where it is possible for others to do so with little cost or risk of harm to themselves.[10]

But if we can only relieve someone's suffering by killing him or by significantly shortening his life, then we have in such cases two rights which conflict, and we can avoid violating one only at the cost of violating the other. Arguing that one should always act so as to err on the side of not violating a fundamental moral right when there is significant danger of doing so is of no help here. In such cases, I believe we should obtain all the available information about the person relevant to what right *he* would choose to waive in such circumstances (including, of course, his views, if they are known, on euthanasia); we must then decide, often under considerable uncertainty, which right it is reasonable to believe he would waive under the circumstances. In such cases, we will have to act on a balance of the evidence, and rarely with the luxury of virtual certainty about whether the person would waive his right not to be killed.

Note that the less information we have about what the specific person in question would choose in the circumstances, the more our reasoning will have to resemble that of what a normal, rational person would choose in the same circumstances. And, in such cases, quality of life considerations will enter into our decisions, since the considerations that would be the basis of the argument that a rational person would desire to die (e.g. that he is suffering from an extremely painful, incurable disease which completely incapacitates him) are quality of life considerations. They show that the quality of life of that person, if his life is not now terminated, will be exceedingly and tragically low. The difficulty with quality of life considerations is not in their entering into any euthanasia decisions at all, but in insuring that they only enter into the question of what a normal, rational person would do in cases where the actual person is unable to waive or decline to waive his right not to be killed, and we do not know what he would do if he could do so. Showing that someone's quality of life is or will be very low can never of itself justify terminating his life.

III

The second type of case of morally permissible termination of human life that I want to discuss is that in which the right to life is overridden by some competing moral consideration. I believe that it is sometimes justified to terminate a person's life against his wishes (self-defense is another important though nonmedical instance already cited), and want to make clear what moral issues are involved and so the kind of reasoning that justifies doing so.

Some of the issues concern the allocation of scarce resources. In this case appeal to cost/benefit analysis is of limited help; perhaps it is better to say that such an appeal presupposes the appropriateness of consequentialist or utilitarian reasoning in an area where it is most controversial and in my view least adequate.[11] What is needed instead is a theory of distributive justice. To see this, first consider the simpler case where only saving lives, i.e. choosing which lives to save, but not improving the quality of lives, is involved. Cost/benefit reasoning would seem to tell us to always save the most lives possible. However, if saving lives is not always to be given priority over improving the quality of lives, then such reasoning will require us to weigh different lives as well according to the expected value they will produce. The life of a great benefactor of mankind, such as a brilliant medical researcher, will count for far more than the life of a good-for-nothing ne'er-do-well. We should save the "more valuable" members of society before and in preference to the "less valuable," and should save a "more valuable" life even in preference to saving several "less valuable" lives if the difference in value is large enough. Many will find this in itself sufficient to discredit utilitarian, cost/benefit reasoning on the issue of killing.[12] However, I do not wish to rest the case against such reasoning on this interpretation, and so shall assume that cost/benefit reasoning requires only weighing lives against lives, with no life to count for more than any other life; it then seems to produce more acceptable results. Where we have sufficient resources to save only some but not all, we should attempt to save the most lives possible. Consider the following:

> *Case* C. We have a limited amount of life-saving medicine needed by two groups of persons. Jones, the only member of group A, needs all the medicine to survive, whereas Brown and Black, the two members of group B, each need only one-half the entire amount to survive.

If we are to save the most lives possible, then we should give the medicine to Brown and Black, thereby saving two lives instead of only one if the medicine is given to Jones.[13] We can say to Jones, "If you consider the situation impartially and imagine that you do not know to which group you belong, then you would agree that we should save Black and Brown in B, for if you don't know to which group you belong, that situation would maximize your own chances of being saved." Or we could say to Jones, "Saving Black and Brown is not unfair to you because if you consider the situation from a fair initial position where you don't know what dosage you will turn out to require and so to which group you will belong, you would accept that we should save Black and Brown." This latter way of putting the argument to Jones is important – it suggests that we want to be able to justify to Jones the fairness to him of our action, and as well that we might not always be able to do so.

Consider another case.

Case D. The situation is identical to that in Case C, except that now Jones is a member of a minority that has been systematically and unjustly exploited by the majority to which Brown and Black belong. It is only as a direct and known result of this exploitation in which Brown and Black have taken part that Jones requires a higher dosage of the medicine for it to be effective.

Were it not for the prior exploitation all three would require equal amounts of the medicine, though there would still be only enough for two, in which case fairness would seem to require a random selection of two among the three to be saved. Thus, if circumstances were not unjust, Jones, as well as Black and Brown, would have an expected probability of being saved of 2/3. As a result of Black's and Brown's exploitation, if we save the most lives possible, Jones has no chance of being saved. Could we still say to Jones that saving the most lives possible is fair or just to him? It seems to me clear that we could not. Rather, Jones could say that to treat him fairly, all three should have an equal probability of being saved, as they would have had in the absence of his unjust treatment at the hands of Black and Brown, which in this case could be achieved by a random selection between saving Jones or saving both Brown and Black. Or, Jones might alternatively argue that he should have the same two-thirds probability of being saved that he would have had in the absence of the unjust exploitation to which he was subject. In real world conditions, of course, things are rarely as clearcut as the example makes them, and so whether simply saving the most lives possible is fair or just to those not saved will in turn often be controversial and unclear. But I would at least suggest that there is a moral requirement of fairness, which is independent of cost/benefit reasoning, and which makes it incumbent on us to explore and consider arguments from fairness in such life-saving situations. I do not believe considerations of fairness should always override considerations of utility maximization, but only that they should sometimes do so; it is difficult to say anything general about how precise trade-offs should be made here. I would add that such life-saving situations occur not only at the level of a physician in possession of scarce life-saving therapy, but as well at policy-making levels in the political process where decisions are made to allocate funds for research or medical delivery systems for different life-threatening diseases.

Next, consider a case where we must weigh saving lives against other benefits, specifically, improving the quality of other persons' lives.

> *Case E.* Suppose rich Jones promises Smith that if Smith will kill Doe, he (Jones) will give a large sum of money to arthritis research. Smith correctly believes that this is the only opportunity in the near future to raise funds for arthritis research and relieve arthritis suffering, and that he can kill Doe with no one else learning of it.

Would Smith be morally justified in killing innocent Doe as the lesser of two evils in order to lessen the suffering and improve the quality of life of arthritis sufferers? Since the distinction between killing and letting die is not, in my view, *in itself* morally significant, it seems that if we simply attempt to maximize benefits and minimize harms we must say that (if Jones's donation is large enough) Smith would be morally justified in killing Doe. Yet cases like this are far more morally troubling than such utilitarian, cost/benefit analysis makes them, and not just for those who mistakenly appeal to the killing/letting die distinction. All can agree that Smith's killing Doe would be *prima facie* wrong because, on the cost/benefit view, it involves the cost of Doe's death, or because, on the rights-based view, it violates Doe's right to life. But is this the only moral consideration to be weighed against the cost of failing to further arthritis research if Doe is not killed? What has been left out again is that it would be unfair or unjust to single out innocent Doe for death to benefit arthritis sufferers. "Why should my life be sacrificed, why me and not others?" Doe might say. Doe is being used merely as a means to improve the lives of others, quite apart from whether *he* deserves to be so disadvantaged for the benefit of arthritis sufferers. A net gain in benefits or reduction of harms or evils from killing Doe, which cost/benefit, utilitarian reasoning might establish, could not show that Smith's action is fair or just *to Doe*. Once again, what is needed here is a comprehensive theory of distributive justice.

Now, just what constitutes an adequate theory of distributive justice is a notoriously complex and controversial issue; it is a major area of contention among moral and political philosophers, as well as the public at large. Tooley's proposal of a fair method for making allocations between lengthening of life and quality of life considerations is therefore, in its simplicity, extremely attractive. He suggests that when everyone has a reasonable personal income, the fair solution is to "allow each person to decide how much of his income he wishes to use, not to improve the quality of his life, but to invest in insurance to guard against the possible occurrences which would necessitate large expenditures in order to keep him alive." This solution is also designed to avoid the well-known problems of the interpersonal comparison of utility and of finding morally defensible methods of social choice or decision.[14]

It will be sufficient to point out two difficulties with this solution to see that it has neither the advantage nor simplicity it promises. First, note that if this is to be a fair solution to the distribution problem, we must stipulate not just that at the outset everyone has a reasonable income, but that everyone has a fair income; only if we begin from a fair status-quo point will the application of fair procedures produce a fair or just outcome. This may seem a quibble, but it is not, for it makes clear that Tooley's proposal does not avoid the need for a theory of the fair or just distribution of income and wealth. And it means as well that so long as substantial injustices persist in the distribution of income and wealth, personal financing of life-saving medical treatment, whether through purchase of insurance or otherwise, will result in substantial injustices in the allocation of such treatments.

Second, setting aside the issue of justice or fairness, we cannot avoid the problems of the interpersonal comparison of utility and of social choice, and so do not have a sound basis for making cost/benefit calculations in the manner Tooley proposes. Successful medical research has some of the attributes of what economists call "public goods"; it makes possible new therapeutic knowledge and procedures which will be generally accessible to members of society. But it is not entirely possible to restrict the benefits of medical research to those who pay for the research when participation in its financing is voluntary, as it would be with Tooley's insurance scheme. Medical research in this respect is similar to such goods as public parks and a system of national defense. There will be a tendency for rational self-interested persons to understate their preferences for public goods, e.g. Tooley's case of arthritis research, because they can benefit from the decisions of others to support the supplying of the goods. This so-called "free-rider" problem in the provision of public goods requires a collectively made and enforced decision to insure that public goods will be provided in amounts reflecting the real preferences of the group. This suggests that a social decision is necessary in order to provide an optimal allocation of funds between life-saving and quality of life programs on cost/benefit grounds. This is a special problem in the present case where we must weigh what appear to be largely incommensurable goods – the saving of some lives versus the improvement in the quality of other lives. Cost/benefit analysis is only of limited usefulness in such decisions.

There is one other common approach to decisions concerning patient treatment that seems to provide a way around some of these conflict-of-interests issues. A common model of the doctor-patient relationship is one in which the doctor is viewed as the agent of the patient and who is to act in the interest and for the welfare of the patient alone. The physician should adopt what has been called a patient-centered ethic.[15]

Decisions about a patient's treatment, including institution or withdrawal of treatment which will hasten or prolong the patient's death, are to be made solely according to how his welfare or interests are affected. This model for treatment decisions has the effect of drastically limiting the conflict of interests category of morally permissible killing. Recall our earlier Case A in which Jones has a terminal disease. He wishes to be kept alive as long as possible, but because of unique features of his medical situation, if he is killed now it is likely that new medical knowledge will be obtained that will enable alleviation of the suffering of other similar patients. The patient-centered model of the doctor-patient relationship would preclude any effects on the interests of persons other than the patient (whether provision of benefits or prevention of harms) from being weighed in the physician's decision about Jones's treatment.

Perhaps more common instances where the interests of

others are significantly affected are those in which continued use of life-sustaining therapies will have disastrous consequences for the psychological and financial well-being of the patient's family. Here again, the patient-centered model of the doctor-patient relationship precludes any consideration of effects on a patient's family in decisions concerning treatment. In terms of a right to life, the patient-centered model gives a particularly strong interpretation to the right to life and indicates when it can be justifiably overridden – it does not allow the conflicting interests of others ever (except perhaps in cases of choosing whom to save and whom to kill or let die) to override the patient's right to life. I am doubtful of how closely physicians in fact adhere to such a model of the doctor-patient relationship in their practice, but, more important, it is by no means clear why they should adhere to it. If there is a class of case involving conflicts of interest in which the termination of human life is morally justified because the effects on the quality of life of others weigh against lengthening the life of the patient, then this doctor-patient model is badly confused and promotes faulty reasoning about an important group of cases.

Where does this leave us with regard to conflict of interests cases? I have suggested that moral considerations of justice and fairness are involved in cases where we must choose which lives to save. And they appear again in cases where saving or lengthening lives must be weighed against improving the quality of lives. Neither Tooley's insurance scheme, nor a patient-centered model of the doctor-patient relationship seems to be a defensible way of meeting these issues. In the moral view I have been elaborating in this paper, the two central issues in the question of when the right to life is overridden by competing considerations remain to be answered. The first concerns the strength or importance of the right to life compared to other moral rights and considerations and to quality of life considerations. The second concerns the fair or just distribution *between* persons of benefits and burdens when saving a life comes into conflict with other considerations. On this second question, I have been content here to argue that utilitarian, cost/benefit reasoning is inadequate, and that the point at which termination of life is morally permissible cannot be settled without some appeal to a theory of distributive justice. Developing such a theory, or even simply applying an already developed theory such as that of John Rawls in his *A Theory of Justice,* is far too complex and lengthy a task to be attempted here. At the very broadest level, the issues include: the allocation of resources between health care and other non-health care interests; the allocation of resources to different sorts of health care needs, for example life-saving versus non-life-threatening; and the extent and basis of any particular individual's legitimate claims on those resources. Unless undue weight is placed on the killing/letting die distinction, we cannot hope to decide when it is morally permissible to kill within the practice of medicine without coming to grips with these broader issues of distributive justice.

With regard to the first issue – the relative importance of the right not to be killed as opposed to other moral rights and considerations – I believe an adequate position will have to account for the special importance normally accorded to the right of life without at the same time according it absolute priority. This special importance reflects, most importantly, two things. First, it reflects the way in which life itself is a necessary condition for carrying out and pursuing all our other life plans and purposes; it is not merely one purpose or project among others, which if taken away could have some other project, with its attendant satisfaction, substituted for it. Second, it reflects the way in which each person's life is uniquely important to him or her and the fact that gains to others in their well-being cannot compensate me, and my life, for its loss; loss of life itself is not able to be compensated for *inter*personally, by benefits gained by others, in the way that satisfactions can be compensated for *intra*personally. The two factors help explain the special importance of the right to life in somewhat different ways. The first factor helps account for the special importance given loss of life in comparison with other harms and benefits, the second for the limits on the possibility of compensating one person for the loss of his life by gains in benefits provided, or harms prevented, to others. Perhaps further development of these and related considerations will yield some insight into the relative strength of the right to life. In the whole category of morally permissible termination of life when the cost of maintaining it is too high, most of the necessary analysis remains to be done; I hope here only to have brought out what some of the issues are, and by so doing to suggest some directions which further inquiry must take.

NOTES

1 Ronald Dworkin, *Taking Rights Seriously*, ch. 6 (Cambridge, Mass.: Harvard University Press, 1977). My characterization of the three types of moral views owes much to Dworkin's discussion.

2 If the moral prohibitions are absolutist prohibitions, then to remain coherent they must be restricted to something like what we deliberately do to others, since one might find oneself in a situation where, whatever one did, the prohibited effect (e.g. someone's death) would come about as a result. Cf. Thomas Nagel, "War and Massacre," *Philosophy and Public Affairs* 1 (1972), pp. 123–44. Neither Nagel, nor anyone else to my knowledge, has made clear precisely how the distinction between what one does, and what merely happens as a result of what one does, should be drawn.

3 This requirement is developed by Nagel, "War and Massacre."

4 See John Rawls, *A Theory of Justice* (Cambridge, Mass.: Harvard University Press, 1971), especially ch. 1 and sec. 87; also Rawls, "The Independence of Moral Theory," *Proceedings of the American Philosophical Association 48* (1974–75), pp. 5–22.

5 There is an important issue that, while not strictly a case of forfeiting rights, is quite closely related – the relevance, if any, to claims for medical care of the patient's own responsibility for his disease or condition. I have in mind cases where the patient could have avoided the condition if he had chosen to;

for example, emphysema acquired as a direct result of heavy smoking.

6   Michael Tooley, "Decisions to Terminate Life and the Concept of a Person," in *Ethical Issues Relating to Life and Death*, ed. John Ladd (Oxford: Oxford University Press, 1979).

7   For a contrary view on personal identity over time, its implications for moral theory in general, and binding our future selves in particular, see Derek Parfit, "Later Selves and Moral Principles," *Philosophy and Personal Relations*, ed. Alan Montefiore (Montreal: McGill-Queen's University Press, 1973).

8   I have discussed paternalism, with specific reference to the involuntary commitment of the mentally ill, more fully in "Involuntary Civil Commitment: The Moral Issues," *Mental Illness: Law and Public Policy*, eds. Baruch Brody and H. Tristram Englehardt, Jr. (Dordrecht, Holland: Reidel, 1980).

9   My colleague Dr. Sidney Cobb has suggested to me that cases of the sort Tooley imagines rarely occur, at least with adults, because loss of cerebral function to the point at which communication becomes impossible is usually associated with relief of pain and suffering.

10 See, for example, David Richards, *A Theory of Reasons for Action* (Oxford: Clarendon Press, 1971), pp. 185–89; and Rawls, *Theory of Justice*, pp. 338f.

11 Precisely how cost/benefit analysis should be understood is not clear. It can be conceived merely as an analytic tool or procedure in which the costs and benefits of alternative courses of action are estimated. I shall associate it as well with a normative principle of a consequentialist or utilitarian sort according to which a course of action ought to be performed if, and only if, it produces at least as good overall consequences (benefits less costs) as alternative possible actions.

12 A good discussion of the implications of utilitarianism for killing in general is contained in Richard Henson, "Utilitarianism and the Wrongness of Killing," *The Philosophical Review* 80 (1971), pp. 320–37.

13 That we need not always save the most lives possible is argued, on somewhat different lines than I have taken, by Elizabeth Anscombe, "Who Is Wronged?" *Oxford Review* (1967), and saving the most lives possible is defended by Gertrude Ezorsky, "How Many Lives Shall We Save?" *Metaphilosophy* 3 (1972), pp. 156–62.

121

14   An excellent discussion of these problems can be found in A.K. Sen, *Collective Choice and Social Welfare* (San Francisco: Holden-Day, 1970).
15   This view is advocated in, among other places, Leon Kass, "Death as an Event: A Commentary on Robert Morison," *Science* 173 (20 August 1971), pp. 698–702.

Chapter 5

Taking human life

It is by now familiar to distinguish moral theories according to whether their basic principle or principles are formulated in terms of goals, rights, or duties. Consequentialist theories are goal-based in this sense, though consequentialists have often argued that nonbasic moral rights or duties can be derived from a basic goal-promotion principle. Within this framework, Alan Donagan's account of common or Judeo-Christian morality is duty-based. He takes the fundamental principle of morality to be, "It is impermissible not to respect every human being, oneself or any other, as a rational creature."[1] First-order precepts are as well duties, specifically, those concerned with the taking of human life and, in particular, absolute duties: "No man may at will kill another," "It is impermissible to kill an innocent human being," or, finally, "It is absolutely impermissible to commit murder."[2] The duty regarding killing concerns both killing oneself as well as killing another, though Donagan specifies several

An earlier version of this article was prepared as an invited address for a conference on the philosophy of Alan Donagan at Illinois State University, November 4–5, 1983. I am grateful to Shelly Kagan, John P. Reeder, and Michael J. Zimmerman for helpful discussion and comments. I am especially indebted to Alan Donagan for his comments prepared for delivery at the conference on the earlier version of this paper, which have forced me to think harder about several matters discussed below. There are a number of important and complex issues addressed on which it is tempting to say a good deal more than I have done here. In revising the article, I have largely resisted that temptation, keeping the focus on my differences with Donagan. I have, however, added a few points in response to Donagan's initial comments when that seemed helpful in sharpening the differences in our views.

qualifications to the prohibition of suicide. I will argue here that, in making the basic moral principle regarding taking human life both a duty and absolute, Donagan is led to an unacceptable moral position regarding the taking of human life. His position is subject to some criticisms commonly made by consequentialists and to others associated with rights-based views, each of which points to what I believe is a more acceptable position on this issue. Since the features of its basic principles defining duties which are also absolute are both more general properties of Donagan's theory and not simply features of his precept regarding killing, the difficulties I suggest may extend to the theory more generally. Moreover, if the absolute duty not to kill an innocent human being necessarily follows, as Donagan believes, from his fundamental principle of morality, then I believe the difficulties that I discuss also give reason to question in turn the acceptability of that fundamental principle, or at least Donagan's interpretation of it. Finally, if there is one common morality, which I doubt, the rights-based criticisms I develop give reason to doubt that common morality is as Donagan describes, while the consequentialist criticisms are more plausibly interpreted as reasons for revising common morality.

I

What, more specifically, is Donagan's position regarding the morality of taking human life? I quoted above the principle, "It is absolutely impermissible to commit murder." And regarding murder, he adds, "Murder is commonly defined as killing the innocent, understanding 'the innocent' in a material sense, as referring to those who are neither attacking other human beings nor have been condemned to death for a crime."[3] As Donagan develops this duty, and as well most others in his theory, the duty holds independent of the wishes of the victim. In particular, it holds whether or not the victim consents to the killing by requesting that the other kill her.[4]

In a certain theological context, this is a natural feature of duties generally. If one's duties are ultimately owed to God, then the fact that the person toward whom one must act wishes one to do otherwise does not extinguish one's duty. God might extinguish one's duty not to kill someone, but the person herself could not do so merely because she wished to die. Her request to be killed would be properly viewed as a request that one act wrongly, a temptation to do evil that should be resisted by a moral person. However, Donagan has sought to develop the Judeo-Christian moral view without appeal to its theological basis, and so it is a Kantian rather than a theological basis that is supposed to justify duties not being limited by the consent of the object of them, of the "victim." I believe the Kantian foundation on which Donagan draws is sufficiently imprecise that it at the least admits of other interpretations that have a rights-based rather than a duty-based character.[5] Rather than engaging in Kantian interpretation, however, I want to say a bit about the rights-based alternative.

If the basic moral principle relevant here defines not a duty but instead a moral right to life which includes a right not to be killed, then the wishes of the victim will be relevant. As H. L. A. Hart has put it, moral rights may be thought of as a piece of moral property of the right holder which makes him a small-scale sovereign in the area of behavior his right covers.[6] In particular, his right gives him control over the obligations of others not to kill him. The right holder can appeal to his right in resisting or protesting any attempt of another to kill him. He can further use the right to extinguish a particular person's obligation not to kill him if he chooses to do so by waiving his right in that instance. (I should emphasize that I am concerned here with moral, not legal, rights. Legal rights not to be killed are not in general waivable, and there are considerations special to the legal process which at least go some way toward justifying this.) Other rights to property, to have promises kept, to privacy, and so forth are commonly understood to be waivable in this way; the correlative obligations they entail are limited by consent, or the waiving of the right, and so of the other's obligation, by the right holder. Either a basic duty not to kill innocent human beings or a basic right not to be killed will serve the end of protecting the important interest persons have in life and in not losing it by being killed by others. If it were always, under any circumstances, in a person's interest not to be killed, then an unwaivable duty not to kill would be well suited to protect that interest. But if there are some circumstances in which continued life is not in a person's interest or reasonably not desired by a person, then a waivable right not to be killed will be better suited to protecting that interest. It will in general protect persons' interest in life while also permitting the right holder to waive his right not to be killed and so extinguish another's duty not to kill him when he judges continued life not to be in his interest. When the basic principle concerning taking life defines a moral right, persons have an additional element of control over the behavior of others affecting their life which they lack if the basic principle defines a nonwaivable moral duty. This is an important element in the value of rights that is lost in a moral system containing nonwaivable duties as basic. In this argument for the rights-based alternative, the appeal is to the nature of persons' actual interest in life and to that interest being better promoted by a rights-based than by a duty-based principle. Some such argument should be possible in moral theories that understand moral limits on behavior as designed to serve the interests of persons, though there is not space here to fill out such arguments.

An alternative line of support for the rights alternative, with the greater control it gives persons, is in terms of its promotion of autonomy, though the proper understanding of autonomy will then be at issue. Kantian theories give a deep place to autonomy, but precisely how autonomy is to be understood is as unclear as how Donagan's Kantian fundamental principle of morality is to be interpreted. There is a plausible sense of autonomy which is related to the specific sort of additional control that rights provide to an individual in comparison with duties and which suggests that rights increase this autonomy in comparison with duties. The autonomy in question concerns the freedom to define for oneself one's own conception of the good. It is not the freedom to define the good for persons generally, that is, the good of all persons, but only one's own good, that is, the ultimate aims, ends, and values that define each person's own view of what is the good life for her. It is also not the freedom to define moral rules or principles of right action governing interpersonal behavior however one wishes. There may be reasonable constraints on an interpersonal conception of the right that do not apply in the definition of one's conception of the good. Finally, within a nonconsequentialist moral theory these principles of right action can constrain the autonomous choice by individuals of their own conception of the good. This is the autonomy associated with liberalism in which principles of right action define the terms of cooperation within which individuals are free to pursue their autonomously defined, different, and conflicting conceptions of the good.

There is then an important reason to prefer a basic moral principle concerning taking human life to be formulated in terms of a right to life, not a duty not to murder, that is, to kill the innocent. But are there important cases in which the rights view has significantly different implications regarding taking life than Donagan's does? Perhaps the place in which questions about taking innocent life are of greatest practical importance is medical care. As medical care has gained an increased capacity to prolong life, the time and circumstances of people's deaths have increasingly become a matter of choice or decision. If health care settings are the place where most decisions to take life are in fact made, then a moral principle regulating such actions ought to have acceptable implications for those decisions and actions. Donagan's principle, however, in my view does not. For virtually all cases of termination of a life-sustaining treatment, the patient in question is "innocent" in Donagan's sense – she is neither an attacker nor condemned to death for a crime. Yet when a physician stops such treatment by turning off a respirator, he performs an action that results in that patient's death. I believe we should say that he thereby kills the patient. Certainly, if the physician stopped the respirator of a patient who could survive indefinitely on that respirator, and without that patient's consent in order to inherit the patient's money, the law, common morality, and, I would expect, Donagan all would concur that he had killed the patient. Suppose for the moment that all actions deliberately stopping a life-sustaining treatment are instances of killing a patient. (I will consider in Section II below some objections Donagan might make to this supposition.) It would seem to follow in Donagan's view that, even when such actions are taken with the free and informed consent of a competent patient, they are always morally wrong. As common morality is expressed in the law, that is not the position of common morality – there is a well-established legal right of any competent patient to refuse any medical treatment, including any life-sustaining treatment, and a legal liberty of the physician to carry out that decision by ceasing treatment. There are various arguments that might be given in support of the law's position here, but I will note only that I believe the fundamental case for it rests in the importance of individual self-determination on so weighty a matter as the time and circumstances of one's death. In sum, Donagan's position is at odds with the law and with common, and in my view justified, moral views, and whether it gives expression to the principle of respecting rational beings as autonomous is also seriously in question.

Many decisions about life-sustaining treatment must be made for patients who clearly are themselves not competent to decide, for example, infants and some seriously debilitated elderly. Here too, the rights position seems more acceptable. Without going into the details, it would direct us to ask whether this person, if competent, would want life-sustaining treatment stopped. In cases in which we lack knowledge of the particular individual's preference or in which the individual never had preferences, we should ask whether any reasonable person would want such treatment stopped or whether stopping treatment is in the patient's best interests. In practice, this will often be very difficult to determine, but in a significant number of cases it will be reasonable to conclude that the decision would be to stop treatment. A rights position can plausibly interpret these cases as extensions for incompetent patients of the waiving of the right not to be killed in the case of competent patients. Assuming appropriate institutional protections against abuse, it has the desirable effect of extending the control individuals can have over decisions about the use of life-sustaining treatment for them to circumstances in which they are no longer competent themselves to decide. Once again, if Donagan's theory condemns as impermissible any stopping by another of life-sustaining treatment of an incompetent patient, its claim to respect the autonomy of rational agents is seriously in question.

Perhaps, however, Donagan's view is closer than it might seem to a position formulated in terms of a basic moral right. I am not confident about the correct answer to this question, principally because of two aspects of his view whose implications he does not fully develop. The first is his discussion of suicide. There Donagan rejects the position "that any human being may quit life as he or she pleases" because, "if one is to respect oneself as a rational creature, one may not hold one's life cheap, as something to be taken at will."[7] But he goes on to sanction suicide in order to spare others excessive burdens or "to obtain release from a life that has become, not merely hard to bear, but utterly dehumanized."[8] Now it might be argued that these exceptions to the suicide prohibition will permit all the stopping of life-sustaining treatment that ought to be permitted. I believe that is doubtful for two reasons.

First, in many instances competent patients decide to stop a life-sustaining treatment not because life has become "utterly dehumanized" or impossible to bear but because its prospects are sufficiently poor to make death welcome. This sometimes occurs in circumstances in which many, or most, other persons would choose continued life. Donagan's examples of a person with hydrophobia and a person trapped in a burning vehicle suggest a very narrow interpretation of when life is sufficiently dehumanized that death may be permissibly chosen. Moral rights, and specifically here a right not to be killed, on the other hand, will give much wider latitude for choice in the exercise of one's rights. They are better suited to reflecting our autonomy interest in defining for ourselves our own conception of the good, including the circumstances in which continued life is, all things considered, a good to a particular person. Whatever the requirement of competence and/or conditions for justified paternalism limiting such choice, they commonly protect persons against some incompetent and harmful choices while leaving them free to make choices according to their own particular aims and values. Another way of putting this point is to say that Donagan holds life under nearly all circumstances to be an objective good and not to be held cheap. In his view, except under the most limited circumstances when life as a rational creature is no longer possible, any choice to end one's life fails to respect oneself as a rational creature. The alternative view I am suggesting takes more seriously our autonomy in defining our own good, as we see it, for ourselves. A moral right not to be killed has the logical structure to give the greater personal control that autonomy so understood supports.

It is worth noting here that Donagan's theory combines two interpretations of the fundamental idea of respecting rational creatures. In the first, rational creatures are to be respected as possessing objective value and so not to be held cheap. In the second, they are to be respected as autonomous, as having the right, subject to the moral law, to decide for themselves what their good is and how to pursue it. The first interpretation leads to Donagan's view that suicide, barring the exceptions he discusses, is wrong, as are self-mutilation, actions impairing one's health, and failure to develop one's mental and physical powers. For Donagan, though he does not put it this way, these are objective features of any person's good, independent of whether the person herself judges them to be good. This first interpretation is in conflict with and puts constraints on a person's autonomy to decide what one's good is and how to pursue it. The rights-based alternative that I am suggesting emphasizes the second interpretation by leaving rational agents free to determine whether and when continued life is for their good.

The second reason for doubting whether the exceptions to the prohibition of killing found in Donagan's discussion of suicide will bring his view closer to what the law and many people's moral views would sanction in medical contexts is simply this. These killings in medical contexts, undertaken with the patient's or surrogate's consent, are often performed by someone other than the patient because the patient is unable to perform the necessary action. If killing another innocent human being is morally impermissible and the other's consent to be killed does not make him noninnocent, then for a physician or family member to stop the patient's life-sustaining treatment is always wrong. The conditions in which suicide is permissible would seem to be beside the point for Donagan when one person kills another. Perhaps then the consent of the patient justifies the physician's acting. To what extent, if any, does consent qualify, for Donagan, the scope of moral prohibitions?[9] Nowhere in the explicit discussion of the prohibition of killing innocent human beings is consent said to limit the prohibition. Moreover, in the discussion of duties to self, as already noted, a person's quitting "life as he or she pleases" or "at will," both of which seem close, if not identical, to "with consent," are explicitly condemned. With one exception, consent and waiver play no significant explicit role in Donagan's elaboration of the theory of common morality. The exception is his discussion of the case, by now well known to philosophers, of the fat potholer stuck in the mouth of a cave, whose companions in the cave will drown in the rising water unless they blast him out with a stick of dynamite. Donagan states:

Human beings are not morally forbidden to risk their lives on suitably serious enterprises, of which it is not unthinkable that

131

potholing is one. And, although it is not beyond dispute, it does not appear to be impermissible for a group of human beings embarking on such an enterprise to agree that if in the course of it, through nobody's fault, they should be confronted with a choice between either allowing certain of their number to be killed, or doing something that would against everybody's will, cause the death of fewer of their number, the latter should be chosen. . . . An agreement of the kind described would have force even if it was tacit: that is, even if, without saying so, all members of the group take it as accepted by all that they should conduct themselves in accordance with it. And perhaps it would have force even if it were virtual: that is, even if all members of the group, were they to think about it, would agree that everybody in the group would think that so to conduct themselves was the only rational course.[10]

But Donagan argues that no such agreement could sanction the potholers knowingly blowing up an innocent picnicker above them in order to blast their way out of the cave. The "doctrine of the legitimacy of consent to possible sacrifice in a common enterprise supplies," Donagan holds, "a ground for the intuitive difference between the two cases."[11] That the appeal to consent as limiting the scope of moral prohibitions is not general but limited to this specific doctrine also helps to explain why Donagan makes no appeal to consent in another case he discusses. Bernard Williams posed the case of Jim the botanist being offered the "privilege" of killing one South American Indian so that the other nineteen rounded up will be let off, whereas if he does not, all twenty will be killed. This also raises the relevance of consent since Williams supposes that all twenty are begging Jim to kill one of them so that the other nineteen will go free. That Donagan does not mention the relevance of the Indians' consent in discussing what Jim may permissibly do suggests the limited scope he is prepared to give to consent as a limitation on moral prohibitions. If morality is conceived as constructed by human beings to in some sense promote and protect their interests while respecting their nature as autonomous agents, then it is reasonable to believe that such autonomous agents would prefer the added control that the rights-based treatment of killing provides, with its general consent limitation on the prohibition of killing, over Donagan's account. I conclude that Donagan's duty-based theory is unacceptable because it fails to give the personal control, and the weight and scope to autonomy, that a rights-based approach to killing provides.

II

The other feature of Donagan's treatment of the morality of killing that I want to focus on is the absolute nature of the prohibition. This too, like the duty-based nature of the view and its limited role for consent, is a more general feature of the overall theory, not just an artifact of his treatment of killing. The difficulties for its absolutist character too go deeper, in my view, than just to the killing prohibition. Donagan holds, as I noted above, that it is always impermissible to murder, that is, to kill the innocent, though he places a limited consent qualification on this duty in the potholer's case. At another place, he states, "In traditional morality, it is unconditionally forbidden (omitting certain minor and contested exceptions) to cause the death of a materially innocent person."[12] Perhaps the most obvious difficulty with this absolutism is that it prohibits killing an innocent person even when doing so is necessary to prevent very great bads, such as the deaths of great numbers of other innocent persons. Donagan defends his position against such "hard cases," though in my view unsuccessfully. However, I am interested here in a less obvious difficulty with his absolutism. While Donagan grants that persons' agency in general extends to what they deliberately omit to do or abstain from doing, he seems to be in accord with much of common morality in drawing a morally significant distinction between acts and omissions. I attribute endorsement of the moral significance of this distinction to him with some hesitation since he does not explicitly address the question. However, it seems clear that the unconditional prohibition is intended to cover killing or causing the death of the innocent but is not extended to not saving the innocent or allowing to die. But this is just to make killing more seriously wrong than allowing to die is. That killing is more seriously wrong than allowing to die is a view shared by absolutists and nonabsolutists alike, but it is a position difficult for any absolutist like Donagan to avoid. If allowing the innocent to die, as well as killing, were unconditionally forbidden, Donagan's theory would almost certainly involve a conflict between absolute duties, and it would as well seem to require persons to devote all their efforts and resources to saving those who would otherwise die.[13] But this would be a stronger requirement to prevent death or to aid others than any branch of common morality accepts. So Donagan's absolutism regarding killing inevitably drives him to assign moral importance to the distinction between killing and allowing to die, that is, between acts and omissions leading to death. I will very briefly argue, following arguments already in the literature,[14] that this distinction is not in itself morally significant and will indicate an important confusion that absolutist views tend to promote, though there is no evidence that this is a confusion to which Donagan himself is subject. In my view, the correct position is that there is no moral difference in itself between killing and allowing to die and that neither is unconditionally forbidden. How might the moral irrelevance of the difference be established? On this, I follow the strategy of arguments advanced by Rachels, Tooley, and others. The strategy consists of constructing a pair of cases, one of which is an instance of killing, the other of allowing to die, but which are similar in all other respects that might be thought to be morally relevant to their evaluation. The argument depends for its force on our concluding that the instance of killing is no more seriously wrong than the instance of allowing to die. The argument depends as well on the claim that, if the properties of being a killing or being an allowing to die are in themselves different in moral importance, then they will have that difference in any pair of cases. And so, if in one pair of cases there is no moral difference, that suffices to establish that in general there is no moral difference between two actions in their degree of wrongness simply because one is a killing and the other an allowing to die. That is, of course, not to say that other properties of actual instances of killings and allowings to die, such as the agent's motive, whether the victim consents, et cetera, may not be morally important and make most actual killings worse, all things considered, than most actual allowings to die. But it is to say that it is then these other properties, and not merely that one is a killing and the other an allowing to die, that are morally important. In order to remain within the medical context and to avoid the bad motives of the agents that in other examples seem to generate some confusion, I offer the following case.

A patient is dying from terminal cancer, undergoing great suffering that cannot be relieved without so sedating him that he is unable to relate in any way to others. He prepared an advance directive at an earlier stage of his disease indicating that these are circumstances in which he wished to have his life ended, either by direct means or by withdrawing life-sustaining treatment. In a recent lucid moment he reaffirmed the directive. The attending physician and the patient's family are in agreement that the patient's desire to die ought now to be granted.

Consider two alternative conclusions to this case:

A. While the patient is deeply sedated and not conscious, his wife places a pillow over his face and he dies peacefully from asphyxiation.
B. While the patient is deeply sedated and not conscious, he develops severe respiratory difficulty requiring emergency professional intervention and use of a respirator; his wife knows this, is present, and decides not to call the professional staff and thereby not to employ the respirator, with the goal of allowing him to die. The patient dies peacefully.

In outcome A the patient is killed by his spouse, while in B his spouse decides not to treat and allows him to die. Is there any reason why what the wife does in A is any different morally than what she does in B? If not, then killing (involving an action leading to death) is in itself no worse than allowing to die (involving an omission leading to death).

Donagan in his initial comments rejects any such argument as obviously logically unsound. It is possible that he understands the claim that there is no moral difference in itself between killing and allowing to die in some different sense than its defenders have intended. I mean the claim to be that the mere fact that one action is an instance of killing and another an instance of allowing to die makes the first no worse (or better) morally than the other and makes no difference in the respective agents' moral responsibility for the deaths. Any moral difference in how seriously wrong the two are, or in the agents' responsibility, must be because of other properties they possess besides one being a killing and the other an allowing to die. Since I believe arguments such as the one I have employed above for this claim are essentially sound, and because they have a larger importance for the killing/allowing to die issue generally, it is worth saying a bit more in defense of such arguments. Why have they seemed to their proponents to be acceptable? I believe the underlying intuition concerns the universality of reasons. Being a killing and being an allowing to die are both descriptive properties of a person's doings (where "doing" is interpreted broadly so as to include omissions). The moral value of such doings depends on, is "supervenient on," as it is sometimes put, their morally significant descriptive properties. Suppose the descriptive properties of being a killing and being an allowing to die contribute some determinate amount to the overall moral values of an actual killing and an actual allowing to die. Then if another pair of doings have the very same properties of one being a killing and the other an allowing to die, mustn't these properties contribute the same amount to the overall value of the actions? If the descriptive properties remain unchanged in the two pairs of cases, and the moral value of the actions depends on these descriptive properties, then how could the moral value of the very same property vary? Why the moral difference between killing and allowing to die should exist in some instances but not in others would appear quite mysterious.

This account makes at least two important assumptions. First, it assumes that it is possible to isolate the moral values or significance of the properties of being a killing and being an allowing to die with a single pair of cases in which no other differences between the two cases are of any moral importance. While it may not be possible to assign determinate numerical moral values to killing and allowing to die in the two cases, we can conclude that the moral values to be assigned to being a killing and to being an allowing to die are the same if the overall values of the actions are judged to be the same. Second, it assumes that the overall moral value of doings is what might be termed an additive function of the values of the distinct and different morally relevant properties or reasons. (If the function was not additive but, e.g., multiplicative, then killing might have a value of -10 and allowing to die -5, yet if other properties were 0, the overall value of a concrete killing and a concrete allowing to die need not differ.) The opponent of this mode of argument should be obliged to show us why the properties are not isolable or why the overall moral value of an action is not an additive function of its different morally relevant properties and, if it is not, what kind of function it is. I believe essentially the same point can be put in terms of reasons for action instead of moral value, where being a killing is understood as a reason for action.

It is logically open to Donagan to say that the killing/allowing to die difference is in itself morally important in some circumstances but not in others.[15] One might hold that when some further property P obtains in cases of killing and allowing to die, but only when P obtains, then a killing will be more seriously wrong, all other things equal, than an allowing to die. No pair of cases such as I and others have offered would be a counterexample to this view unless P obtains in each case in the pair. I am unclear whether this is Donagan's view, but if it is, we must be told what P is in order to see if any pair of cases can be constructed as a counterexample against the putative moral importance of the killing/allowing to die distinction when P obtains. It should be emphasized that we can then employ the very same strategy of argument that Donagan rejects, only now being sure that P obtains in each case. We also will want some explanation of why the killing/allowing to die difference becomes important when, but only when, P obtains. If this is Donagan's view, I am unable to evaluate it because it is not clear to me what he believes P to be. And if this is Donagan's view, should we say that in it the mere difference between two actions, that one is a killing and the other an allowing to die, is in itself morally important? I believe the answer would be no in one sense and yes in another: no because the moral difference between killing and allowing to die only exists when the additional property P obtains; yes because when P does obtain in each instance it is whether a doing is killing or allowing to die that is morally important.

Let us look more explicitly at what Donagan takes to be the standard defense of the difference between killing and allowing to die. That is that moral agents are responsible for their own actions and inactions but not for the course of events except inasmuch as some permissible action in their power would have made a difference to it. When an evil can be prevented only by an impermissible action, there is a morally significant difference between allowing it to happen and causing it. The former is right, the latter wrong. I believe this defense fails because it begs exactly the question at issue. The structure of the cases would seem to be this:

1. I can cause an evil E.
2. I can do A, which will prevent an evil E.

But Donagan argues that, if A is impermissible, I must allow E to happen in 2, though in 1 I must not cause E. However, we must look more carefully at the structure of the cases:

1. I can either *(a)* do A', thereby causing E, or
   (b) omit A' and not cause E.
2. I can either *(a)* omit A, thereby allowing E to happen, or
   (b) do A and thereby prevent E.

Donagan seems to assume that doing A in 2 is impermissible but that omitting doing A' in 1 cannot be. And that is what is question begging. In 2, doing A has the good effect of preventing E but is impermissible for some other reason (e.g., because it is also a killing). But then in 1, why can't omitting A', though it has the good result of not causing E, also be impermissible for some other reason (e.g., it is also an allowing to die). The cases can be directly analogous and so cannot be offered as a reason why killing can be in itself worse than allowing to die. For the cases to work as Donagan supposes, it must already have been established on other grounds that killing is in itself worse than allowing to die or is in itself impermissible while allowing to die is not. The other grounds one would expect Donagan to appeal to would ultimately be some difference in the relative respect displayed for rational creatures. But why should killing someone display less respect for them than allowing them to die, when all other circumstances are equal? What Donagan's nonconsequentialism may require is that, when an evil can be prevented only by doing something impermissible, then it is wrong to prevent it. But an impermissible doing might be either an act or an omission, either killing or allowing to die. The fact that one cannot do the impermissible that good may come or evil be prevented provides no grounds for thinking that only killing but not allowing to die is impermissible.

I believe it is also open to argue here against Donagan that, even should (contrary to what I have argued above) saving sometimes be wrong when in an otherwise similar case not killing would be right, the difference being that saving involves doing something impermissible for other reasons, this does not show that killing is ever in itself morally different than allowing to die. That something impermissible might in some cases be required to save would seem to be some further property of the actual instance of saving beyond it merely being a case of saving. It cannot be an essential property of saving since not all instances of saving require doing something impermissible.

Since arguments for the moral irrelevance of the difference between killing and allowing to die are often made by proponents of consequentialist moral theories, I would emphasize that accepting those arguments does not commit one to consequentialism. But accepting them does imply that a basic moral principle evaluating taking life should not, as Donagan's does, mark a moral difference between killing and allowing to die. There is much at stake for a moral theory like Donagan's (though also for moral theory generally) in this issue. As already noted, if there is no moral difference in itself between killing and allowing to die, then absolutism with regard to the prohibition will be impossible to maintain. On the other hand, if the killing prohibition is to remain extremely strong, even though not absolute, and one must not kill even at great cost to oneself or others, then one must not allow to die, other things being equal, but instead save even at similar cost to oneself or others. A very strong principle requiring lifesaving aid may be the implication of a very strong prohibition of killing. But the implications for a principle of aid to others are not what I want to explore here.

Rather, consider again a standard sort of case of termination of life in the medical context – stopping life-sustaining treatment, such as by turning off a respirator. I argued in Section I that doing so seemed to be prohibited by Donagan's prohibition of killing the innocent and that this is an unacceptable result. Donagan might reply that such acts are not absolutely prohibited in his theory because they are not cases of killing but instead instances of allowing the patient to die and allowing to die is not absolutely prohibited. I am unsure whether Donagan would make this reply, but it is often the reply of upholders of common morality. Such a reply is at least helpful in maintaining the plausibility of an absolute prohibition of killing. But I believe, nevertheless, that it is mistaken. Consider again the case of a disabled but competent patient, able to survive for months, perhaps years, on a respirator, but who comes to a firm conclusion that he wants the respirator stopped and to die. The physician accedes to his request and turns off the respirator. Why is it so commonly said, by physicians and the general public, that this is merely a case of allowing the patient to die? One reason is that in similar cases in which the patient is dying imminently many believe that the physician's action merely accelerates or does not prolong the dying process. It is the underlying disease, it is said, not the physician's action, that causes death or kills. Common though this way of thinking is, I believe that it is confused. The notions of "dying process" and "causality" employed here have never, in my view, been clarified in a manner that allows them to sustain the moral weight being placed on them. However, since Donagan offers no explicit account along these lines, I do not evaluate it further here.

A second important motive for construing such cases as allowing to die is that common morality is often understood to condemn all killing of the innocent, yet this case seems to many to involve both an innocent patient and a morally acceptable action. Rather than qualify the scope of the prohibition of killing with a consent limitation, in my view the correct move, the case is instead interpreted as not an instance of killing. I believe that is a mistake not only because confused but also because, though it helps allow killing the innocent to continue to be viewed as always wrong, such cases remain fatally vulnerable to the charge that they are killings and therefore wrong.[16] Imagine that, before the physician could act, the patient's nephew, tired of "waiting for the old man to die" to inherit his money, comes into the room and turns off the respirator causing his uncle to suffer respiratory distress and then death. The nephew would be charged with murder in the law and condemned morally for wrongful killing. A response by the nephew that he did not kill his uncle but merely allowed him to die or let nature take its course and that his respiratory illness killed him would quite rightly be treated as a piece of specious absurdity. Yet while the nephew's motive for killing is different than the physician's, the act he performs in turning off the respirator, which leads to the patient's death, seems to be no different than what the physician does. The difference between killing and allowing to die is itself not a difference in the motives of the agents. Because the physician acts with the patient's consent and with a benevolent motive, what he does is evaluated morally as different from what the nephew does who acts from a bad motive. That is not to say that only the nephew kills, however, but instead that only the nephew kills wrongly while the physician kills with justification. The tradition of common morality, which Donagan supports, that views killing the innocent as unconditionally prohibited, plays a powerful role in reinforcing and maintaining confusion about these cases, though I would repeat that Donagan himself cannot be accused of this confusion.

NOTES

1   Alan Donagan, *The Theory of Morality* (Chicago: University of Chicago Press, 1977), p. 66.
2   Ibid., pp. 83, 87–88, 88.
3   Ibid., p. 87.
4   But see the discussion below of consent for one qualification of this claim.
5   For example, both John Rawls, *A Theory of Justice* (Cambridge, Mass.: Harvard University Press, 1971), and Charles Fried, *Right and Wrong* (Cambridge, Mass.: Harvard University Press, 1978) develop rights-based theories with explicitly Kantian underpinnings.
6   Hart develops this view of rights in many places in his writings. See esp. "Are There Any Natural Rights?" and "Bentham on Legal Rights," in *Rights*, ed. David Lyons (Belmont, Calif.: Wadsworth Publishing Co., 1979), pp. 14–25, 125–48.
7   Donagan, p. 76.
8   Ibid., pp. 78–79.
9   I have discussed the relation of consent to absolute moral prohibitions in more detail in "Moral Prohibitions and Consent," in *Action and Responsibility*, ed. M. Bradie and M. Brand, Bowling Green Studies in Applied Philosophy (Bowling Green,

Ohio: Bowling Green State University, Applied Philosophy Program, 1980).

10 Donagan, pp. 178–79.

11 Ibid., p. 180.

12 Ibid., p. 51.

13 I believe the theory even as it stands can yield a conflict between absolute duties. To adapt a case of Judith Thomson's ("Killing, Letting Die, and the Trolley Problem," *Monist* 59 [1975]: 204–18), suppose you are the driver of a trolley on which the brakes have failed. There is a fork in the track just ahead, and you can steer to the right killing five workmen on the right track or to the left killing one workman on the left track. I would want to say in this case that, whichever way you steer, you kill someone. Thus, even if the duty is restricted to not killing I believe there is reason not to conceive it as absolute in order to avoid conflicts of absolute duties in the absence of prior wrongdoing.

14 James Rachels, "Active and Passive Euthanasia," *New England Journal of Medicine* 292, no. 2 (January 9, 1975): 75–80; and Michael Tooley, "A Defense of Abortion and Infanticide," in *The Problem of Abortion*, ed. J. Feinberg (Belmont, Calif.: Wadsworth Publishing Co., 1973), pp. 84–86.

15 Jonathan Bennett explores this possibility in more detail in "Positive and Negative Relevance," *American Philosophical Quarterly* 20, no. 2 (1983): 185–194.

16 Jonathan Bennett has made the interesting suggestion that this is a case both of killing and of allowing to die, interpreting the distinction as partially overlapping rather than as mutually exclusive. If this is so, it is still compatible with my view that this is not a case merely of allowing to die but a case either as well or instead of a killing. See Jonathan Bennett, "Morality and Consequences," in *The Tanner Lectures on Human Values*, vol. 2, ed. S. M. McMurrin (Salt Lake City: University of Utah Press, 1981), pp. 45–116.

Chapter 6

Death and dying

In recent decades medicine has gained dramatic new abilities to prolong life. Patients with kidney failure can be placed on renal dialysis; patients who have suffered cardiac arrest can sometimes be revived with advanced life-support measures including drugs, electric shock, airway intubation and closed or open heart massage; patients with pulmonary disease can be assisted by mechanical ventilation on respirators; and patients unable to eat or drink can receive nourishment and fluids intravenously or with tube feedings. These are only some of the most dramatic and well-known additions to medicine's armamentarium for staving off death in the gravely ill. While these and other life-sustaining treatments often provide very great benefits to individual patients by restoring or prolonging functioning lives, they also have the capacity to prolong patients' lives beyond the point at which they desire continued life support or are reasonably thought to be benefitted by it. Thus, where once nature took its course and pneumonia was the "old man's friend," now increasingly someone must decide how long a life will be prolonged and when death will come. This chapter addresses some of the principal moral issues and arguments in current debates about life support. First, I address very briefly a related issue: the definition of death.

Death and dying

The traditional criterion for determining death until recent years was the permanent ceasing to function of the heart and lungs. When a person stopped breathing and his or her heart stopped beating for more than a few minutes the loss of function was irreversible and the patient was declared dead. The loss of oxygen to the brain would quickly produce irreversible brain damage and loss of all cognitive function. However, the advent in recent years of new medical technology, and most importantly of respirators, has enabled modern medicine to continue artificially patients' heart and lung function after they would no longer function unassisted. As already noted, this can often save lives that previously would have been lost and sometimes permit the patient to recover a normal level of functioning. In some few other cases, however, heart and lung function can be restored or continued by these artificial means after brain function has been partially or completely destroyed, for example, from prolonged loss of oxygen or severe trauma to the brain. Such cases have forced a rethinking of the criteria for the determination of death (President's Commission 1981).

This rethinking has led to a widely acknowledged additional criterion for death, the complete and irreversible loss of all brain function, or so-called brain death. This allows a patient to be declared dead who has suffered complete and irreversible loss of brain function even if the patient's respiration and circulation are being continued by artificial means. There remains a single unitary *concept* of death, for example, "permanent cessation of the integrated functioning of the organism as a whole," or "irreversible loss of personhood" (Capron and Kass 1972, 102–104). What has been introduced is an additional criterion for when this condition of death obtains, namely the complete and irreversible loss of all brain function. Of course, once this additional criterion has been accepted, the practical question then is precisely which medical tests and procedures will establish this loss of brain function. This is a matter for medical determination given a particular level of medical knowledge and technology; it will change over time and will not be explored here (President's Commission 1981, Appendix F). However, I will consider the principal area of philosophical controversy between so-called whole brain and higher brain formulations in the new criteria for the determination of death.

The new definitions adopted by state courts and legislatures, as well as by various official bodies that have studied the matter such as the President's Commission, have employed the whole brain formulation. In this formulation it is the loss of functioning of the whole brain, either as the integrating mechanism of the body's major organ systems or as the hallmark of life itself of the human organism, that is required for death. To quote from the President's Commission:

> When all brain processes cease, the patient loses two important sets of functions. One set encompasses the integrating and coordinating functions, carried out principally but not exclusively by the cerebellum and brainstem. The other set includes the psychological functions which make consciousness, thought, and feeling possible. These latter functions are located primarily but not exclusively in the cerebrum, especially the neocortex. (President's Commission 1981, 38)

Thus, the whole brain formulation includes both the loss of integrating functions that make natural respiration and circulation possible as well as the functions that make consciousness, thought, and feeling possible. The higher brain formulations focus exclusively on the latter functions. In this view, it is consciousness, thought, and feeling that are necessary to personhood, and when they have been irreversibly lost the *person* has died or permanently ceased to exist (Green and Wikler 1980; Veatch 1975). Anyone dead by the whole brain formulation will obviously be dead as well by the higher brain formulation, but not vice versa.

In particular, brain injury resulting from stroke or trauma may permanently destroy all capacity for consciousness, thought, and feeling, but may allow respiration and circulation to continue either assisted or unassisted. This is essentially the condition of patients in so-called persistent vegetative states such as Karen Quinlan, who was able to breathe on her own for a number of years before she died.

There is not space to explore here deep philosophical questions about the nature of personhood and personal identity that divide whole and higher brain formulations. Some version of the higher brain formulation is probably supported by most accounts of personhood, but also at issue is whether the appropriate criterion is the death of a person, which would seem to be the issue of moral concern, or the death of a human being or organism. Since the determination of death is principally a legal determination, practical considerations of social and legal policy are also relevant. Use of the higher brain conception requires the willingness to declare as dead persons whose body's circulatory and respiratory functions remain intact; that is, those whose bodies are still breathing on their own. It would also require practical methods of discrimination of brain function that permit determinations of the permanent loss of higher brain function while some lower brain function persists, and to do so with the very high degree of certainty reasonably required for any declaration of death.

Since public and legal policy has for the present largely settled on the whole brain conception, we shall assume the perspective of that conception in the discussion that follows. It is to be emphasized that on that conception a patient such as Karen Quinlan who has suffered a permanent loss of consciousness and of all capacity for thought and feeling has *not* died. Such patients may, however, constitute a unique class of patients for decisions about forgoing life support because the permanent loss of all consciousness may imply the lack of any possible interest in continued life.

AN ETHICAL FRAMEWORK FOR
LIFE-SUPPORT DECISIONS

Any account of morally appropriate procedures and content for life-support decisions presupposes a broader account of medical treatment decisionmaking generally, and of the physician/patient relationship. These broader issues are addressed more fully in Chapter 1 of this volume, on informed consent. Here, I will set out only briefly the issues in a general account of treatment decisionmaking and the physician/patient relation to guide life-support decisions. It is all too easy to exaggerate the shift in actual practice regarding medical treatment decisionmaking that has taken place in recent decades, and to reduce discussions of models of physician/patient relations to caricatures. With that said, however, it is widely agreed that it was common historically to view the physician/patient relation as one in which the physician directed care and made decisions about treatment, while the patient's role was to comply with the "Doctor's orders." Patients were told only as much about their condition and treatment as was necessary to comply effectively with treatment.

This is sometimes called the authoritarian or paternalist model of the physician/patient relation. It reflects the medical training, knowledge, and experience possessed by the physician but lacked by the average patient, as well as the anxiety, fear, dependency and regression common in critically ill or dying patients. Moreover, if the end of medicine is seen as the preservation of the patient's health and life by the treatment of disease, it is not then surprising that physicians are viewed as possessing the necessary expertise to determine which treatment will best do so. With the dramatic increases in recent decades in medical knowledge and expertise, as well as the many new modes of treatment now possible, the case for the physician as primary treatment decisionmaker now would seem all the stronger. Yet the weight of argument and opinion has shifted substantially toward securing an enlarged, indeed principal, role in treatment decisionmaking for the patient (President's Commission 1982). Why has this happened? There are, of course, many historical reasons for this shift that I will not detail here, but there has also been a reconception of the ends of medicine and of the proper form of the physician/patient relation (Katz 1984; President's Commission 1982; Siegler 1981).

There are many ways of more precisely formulating this newer conception. One prominent version sees the goals of health-care decisionmaking as the promotion of patients' well-being while respecting their self-determination (President's Commission 1982, Ch. 2). How is this different from promoting and preserving patients' health and life? What will best promote health and life is naturally thought to be an objective factual matter, an empirical question, and not a function of individual preference and value. So understood, what will best promote patients' health and preserve their lives is a factual matter about which the physician, not the patient, possesses expertise. Why then is a central role necessary for the patient in selecting the treatment best for the patient?

To conceive the endpoint of the health-care process as the patient's well-being, instead of as health and life in general, is not to deny that physicians seek beneficially to affect patients' health and life. Rather, it is to stress that health and life extension are ultimately of value in the service of the broader overall well-being of the patient. They are of value in so far as they facilitate the patient's pursuit of his or her overall plan of life; the aims, goals and values important to the particular patient. In many instances the decision of which alternative treatment will best promote a patient's well-being, including the alternative of no treatment, cannot be objectively determined independent of that patient's own preferences and values.

In the case of life-support decisions, when the forgoing of life support is under serious consideration it is usually because (1) the patient is critically or terminally ill and likely to die soon no matter what is done, and (2) the quality of the patient's life is seriously limited by the effects of disease, disability and sometimes the treatment itself. Whether treatment and continued life under such severely constrained conditions are better than no life at all must depend in significant part on how the particular patient views his or her life under those conditions. While the physician will be in the best position to predict the specific outcomes of different treatment alternatives and their effects for the patient, the patient will be in the best position to evaluate what importance should be given to any particular effect, such as the discomfort and restrictions on communication of intubation, or the restrictions on activities of dialysis. It is well established that different persons evaluate the importance of such burdens significantly differently – some will tolerate intubation or dialysis relatively well if it allows their lives to be prolonged, while others find that the limitations make life no longer worth living.

There is no single right answer to how such conditions should be valued; there are only the actual answers that real persons give for themselves. This is why many have urged that health-care decisionmaking should be a process of shared decisionmaking between physician and patient (Katz 1984; President's Commission 1982; Siegler 1981). Each brings something to the decisionmaking process that the other lacks, and the communication is necessary to decisions that best serve the patient's well-being. The physician brings knowledge about the likely outcomes of alternative treatments; the patient brings knowledge of the personal aims, ends and values by which to evaluate those outcomes. Thus, even if treatment decisionmaking aims only best to serve the patient's well-being, shared decisionmaking, a process of conversation between the physician and patient, is necessary to identify the best alternative.

In this more recent view of health-care decisionmaking, the other value that should guide the process is the patient's self-determination or autonomy. Self-determination can be understood as the interest each person has in making important decisions that shape and affect one's life for oneself and in accordance with one's own aims and values (Dworkin

1988). Self-determination reflects the everyday importance people give to having control over the course of their lives and thereby the level of responsibility they require to direct their lives. The claim of a right to self-determination is the claim of a right to define and revise one's desires and values in light of experience, and to pursue one's own conception of a good life. Involving patients in important treatment decisions, and leaving them free to refuse any proffered treatment, respects their self-determination. If the interest in self-determination is important in non-life-support cases that affect patients' lives in significant ways, then surely it is at least as important in decisions concerning whether treatment to sustain their lives will be employed or forgone, and thus when and under what conditions their lives will end. Valuing self-determination requires respecting both patients' own conception of their well-being (the subjective aspect of well-being noted above) as well as patients' interest in participating in the decisionmaking process about their care.

Both these values of patient well-being and self-determination support a process of shared decisionmaking between physician and patient in which the patient retains the right to refuse any offered treatment. Shared decisionmaking does not preclude, however, the important trust traditionally and commonly bestowed by patients on their physicians, even when patients ask their physicians to decide for them. There is a crucial moral difference, however, between patients transferring their right to decide to a trusted physician or family member and patients deferring to physicians because decisionmaking is considered to be physicians', not patients', business. The ultimate right of refusal of treatment rests with the patient because it is to or on the patient that treatment is done. It is the patient's body, and in turn the patient's life, that bears the principal effects of any treatment instituted.

The increased role for the patient in a process of shared decisionmaking has not, of course, gone unchallenged and unprotested (Kass 1985; Sider and Clements 1985). Some commentators have insisted that the proper goal of medicine is the objective (as opposed to subjective) promotion of health; for example, health defined in terms of normal species function. They argue that physicians are in the best position to determine what will best serve this goal and that physicians should not be guided solely by the patient's preferences. Many physicians no doubt also share this resistance to any incursion on their traditionally dominant decision-making role. Nevertheless, the debates generally have not made clear to what extent, if any, these commentators would limit a competent patient's right to refuse any life-sustaining treatment. Disagreement probably more often concerns whether a physician must accede to a request, either from a competent patient or more likely from an incompetent patient's family, for life-sustaining treatment that the physician believes is inappropriate. All generally agree that no physician should be required or forced to provide treatment that he or she believes is not within the bounds of acceptable medical practice, but a substantive difference may remain about how these boundaries should in a particular case be defined and how responsive they should be to patient or family preferences.

The account of decisionmaking about life-sustaining treatment based on the values of patient self-determination and well-being empowers the competent patient, or the incompetent patient's surrogate, to weigh the benefits and burdens of alternative treatments, including the alternative of no treatment, according to the patient's own view of the relative value and importance of the various features of those alternatives (President's Commission 1983a). If the competent patient, or the incompetent patient's surrogate, judges that the overall benefits and burdens of the alternative employing life support are worse for the patient than the alternative of forgoing treatment with its expectation of death, that choice is to be respected. It essentially involves a judgment by the patient or surrogate that the expected duration and quality of the continued life possible for the patient is so bad that on balance it is worse than no further life at all.

It should be noted that in this view so-called quality of life considerations *do* quite properly and inevitably play a role in the assessment of alternatives and of their overall benefits and burdens. For most persons, whether the continued life made possible by a particular life-sustaining treatment is on balance a benefit to them will depend at least in part on the quality of that life. What is important to emphasize is that the assessment should be of the quality of life *to the patient* and whether its quality makes continued life on balance a benefit or burden to him or her. This view does not sanction giving weight to how the patient's continued life may affect the quality of others' lives, for example by making the patient a burden to others. Nor does it sanction any judgments that some people's lives are not socially or economically worth sustaining because they are of low quality. The proper question is whether the patient's present and anticipated quality of life is sufficiently bad to make it, according to him or her, worse than no more life at all. This is a very narrowly constrained role for quality of life considerations that is fully compatible with respecting patients' self-determination and their own view of their well-being.

THE INCOMPETENT PATIENT

This account of life-support decisionmaking thus far largely assumes that the patient is competent to make such decisions. Of course, this is often or even usually not the case when forgoing life-sustaining treatment is seriously at issue. The effects of illness and disease, as well as of treatments themselves, commonly compromise or eliminate patients' abilities to participate in decisionmaking. Someone else will then have to decide for them. This chapter is concerned with moral issues concerning life support and so I will not discuss in any detail the various legal and institutional practices and procedures that have been developed for decisionmaking for incompetent patients (President's Commission 1983a, Ch. 4, Appendices D, E, F). However, I will briefly discuss the moral principles that can guide decisionmaking for incompetent patients (Buchanan and Brock 1986).

Probably the most direct way for incompetent patients to participate in decisions about their care in a manner serving their well-being and self-determination is through use of advance directives. Advance directives, the best known of which are so-called Living Wills, enable persons while competent to specify their wishes about treatment should they later become incompetent. A majority of the states in the United States now have given these documents legally enforceable status. Nevertheless, advance directives are at best only a partial solution to the problem of decisionmaking for incompetent patients for several reasons. First, and probably most important, only a very small proportion of incompetent patients for whom such decisions must be made have ever given advance directives. This is likely to continue to be the case even with increased efforts to publicize advance directives and their value.

Second, to ensure the patient's competence when they are given, advance directives are usually made well in advance of the circumstances in which they are to be applied. Thus, they are inevitably framed in somewhat vague and general terms, and commonly make use of phrases like "if I am terminally ill and death is imminent, no further artificial or extraordinary means to prolong my life shall be employed," and so forth. While such instructions can provide others with general guidance as to the patient's wishes regarding life support, they inevitably leave much discretion to those who must interpret them in the patient's specific circumstances. At what point is death imminent? Are antibiotics extraordinary means? Even when advance directives have been given, others must unavoidably play a role in decisionmaking about life support for incompetent patients to interpret those directives. This has led many persons to conclude that so-called Durable Powers of Attorney for Health Care, by which persons designate who is to make treatment decisions for them if they later become incompetent, are more helpful than Living Wills.

A third difficulty is that in order to guard against possible well-intentioned misuse or ill-intentioned abuse by others of advance directives the conditions bringing the directives into legal effect are often narrowly limited. For example, the condition that death be imminent on many natural interpretations restricts the directive so that it does not apply in many of the circumstances in which decisions about life support must be made. In any event, it is rarely the case that such directives are taken to court to enforce action in accordance with them. Instead, their function has been to serve in a more informal way as evidence of what the patient would have wanted in the circumstances.

In the absence of any advance directive, others must decide for the incompetent patient. The principle guiding such decisions most in accord with promoting the patient's well-being, as he or she conceives it, while also respecting his or her self-determination, is the principle of Substituted Judgment (Buchanan and Brock 1986; Capron 1982). This principle directs the surrogate decisionmaker to attempt to decide as the patient would have decided if he or she had been competent. This essentially directs the surrogate to take account of any available information about the patient's preferences and values relevant to this decision, even if these preferences and values are different from most people's, in determining what the patient would have wanted.

In the absence of any information about what the particular patient would have wanted, for example because there are no available family or friends of the patient, it is generally accepted that the principle guiding decisions should be the Best Interest Principle. This principle directs the surrogate to decide about life support in a manner which best serves the patient's interests. Lacking any knowledge of this particular patient's wishes, such decisions will inevitably involve asking what most reasonable persons would want for themselves in the circumstances.

It is widely agreed that the surrogate decisionmaker who is to apply these principles should usually be the patient's closest family member. The presumption for the family member as surrogate is usually based on four reasons. First, a family member will usually be whom the patient would have wanted to serve as surrogate. Second, in most instances the family member will know the patient best and so be in the best position to determine what the patient would have wanted. Third, in most cases the family member will care most about the patient's well-being and so be most concerned to represent it. Finally, the family in our society is commonly accorded a significant degree of decisionmaking authority and discretion concerning its members in order to preserve the value of the family.

It should be emphasized that these considerations only establish a presumption for the family member as surrogate. They do not imply that the family member is always the appropriate decisionmaker, but only that in most cases a family member will be a better surrogate than anyone else who as a general practice might be turned to. When these reasons supporting the family member as surrogate do not hold in a particular case, for example because there is an evident conflict of interest between the patient and the family member or because the patient and family member have been estranged for many years, someone else should serve as surrogate. The physician may then have a positive obligation to ensure that the family member is removed as surrogate, through appeal to the courts if necessary.

There are a number of points of controversy concerning surrogate decisionmaking that I shall merely note here. One concerns the proper procedures and standards for determining whether the patient is incompetent to decide for him- or herself. In cases of questionable competence, this is a complicated matter that is dealt with in more detail in Chapter 1, "Informed Consent." Very roughly, what is needed by a patient is adequate capacity to understand relevant information about alternatives and their consequences, together with the ability to apply one's own values to those alternatives and to select one as best (Buchanan and Brock 1986; Drane 1985). A second point of controversy is whether a surrogate should have the same range of discretion in choice that a competent patient would have. Third, in what cases and to what extent should it be required to involve others in the surrogate decisionmaking process? For example, when and how might institutional ethics committees within hospitals and other health-care delivery institutions become involved (Fost and Cranford 1985; Rosner 1985), and when is court review of decisions desirable? Despite these and other areas of controversy about surrogate decisionmaking for incompetent patients, however, the fundamental ethical framework discussed above for the competent patient that appeals to the values of patient well-being and self-determination can be extended to the incompetent patient as well.

SOME ADDITIONAL CONTROVERSIAL MORAL CONSTRAINTS ON FORGOING LIFE SUPPORT

Many of the principal moral disputes about life-sustaining treatment do not focus on the broad issues discussed above of the proper role of the patient in health-care decisionmaking and of the proper form of the physician/patient relationship. Instead, these disputes are more specific to life-sustaining treatment, reflecting the important fact that death is the direct and expected result of forgoing such treatment, and have taken several forms. First, are there special constraints on what is morally permissible regarding life support because life and death are directly in the balance? For example, while it may be morally permissible not to start a particular life-sustaining treatment, is it also permissible to stop the treatment once it has begun? Would doing so be to kill the patient, not merely to allow him to die, and if so would it therefore be wrong? If life-sustaining treatment can be withdrawn with the expectation that death will result, is euthanasia permissible – for example, directly killing, as by a lethal injection, a terminally ill and suffering patient who voluntarily requests it?

These are only some of the special issues and distinctions that I shall consider below and that are important in the debates about moral limits on acceptable action concerning life-sustaining treatment because life and death are in question. Since the prohibition of the intentional killing of an innocent human being is one of our society's strongest moral and legal norms, it is hardly surprising that these issues should be difficult and controversial.

A second area of substantial recent concern is whether there are morally important differences between different kinds of treatment that would make forgoing some treatments impermissible in circumstances in which forgoing others would be permissible. For example, some persons believe food and water should never be withheld though treatments like mechanical ventilation and dialysis may be. Third, many persons find morally important differences between different kinds of patients, justifying either greater restrictions on withholding life support, or greater latitude in doing so. Examples of each sort include critically ill newborns and patients in a persistent vegetative state. In the remainder of this chapter, I shall work somewhat systematically through these moral issues in the care of the dying.

One issue that will not be addressed below is the economic costs of care of the dying. Whether these costs are excessive is controversial (Bayer et al. 1983). The precise nature, basis and scope of a general social obligation to ensure access for all to an adequate level of health care, including life-sustaining care, is also controversial and complex (President's Commission 1983b). These issues are beyond the scope of this chapter, but are addressed in several of the chapters in Part III. However, since some 30 million Americans are substantially without access to health care, I will assume here that ability to pay should not be a general moral criterion limiting access to life-sustaining care.

Withholding and withdrawing life support

One constraint that many persons accept on a patient's right to forgo any life-sustaining treatment judged to be excessively burdensome is that while it can be morally permissible not to start a particular life-sustaining treatment, it is not permissible to stop it once it has begun. Alternatively, even if such treatments can sometimes be stopped once begun, it is often held that it is a graver matter requiring weightier reasons to stop; stopping is at least sometimes not permissible in circumstances in which it would be permissible not to start. This accurately reflects much medical practice in which, for example, physicians who are prepared to honor patients' or their families' requests for "Do Not Resuscitate" or "Do Not Intubate" orders nevertheless in similar circumstances will not stop respirators on which patients are dependent for life. Physicians commonly feel more responsible for a patient's death that results from stopping a respirator than from not starting one. But is there good reason to treat withdrawal of life-sustaining treatment as morally different and more serious than withholding such treatment? Consider the following case.

A very gravely ill patient comes into a hospital emergency room from a nursing home and is sent up to the intensive care unit. The patient begins to develop respiratory failure that is likely to require intubation very soon. At that point the patient's family members and long-standing attending physician arrive at the ICU and inform the ICU staff that there had been extensive discussion about future care with the patient when he was unquestionably competent. Given his grave and terminal illness, as well as his state of debilitation, the patient had firmly rejected being placed on a respirator under any circumstances, and the family and physician produce the patient's Living Will to that effect. Most would hold that this patient should not be intubated and placed on a respirator against his will. Suppose now that the situation is exactly the same except that the attending physician and family are slightly delayed in traffic and arrive fifteen minutes later just after the patient has been intubated and placed on the respirator. Can this difference be of any moral importance? Could it possibly justify morally a refusal by the staff to remove the patient from the respirator? Do not the very same circumstances that justified not placing the patient on the respirator now justify taking him off it? Do not factors like the patient's condition, prognosis, and firmly expressed competent wishes morally determine what should be done, not whether we do not start, or fifteen minutes later stop, the respirator? Why should the stop/not start difference matter morally at all?

Cases such as this have led many to conclude that the difference between not starting and stopping, or withholding and withdrawing, life-sustaining treatment is not in itself of any moral importance (President's Commission 1983a; Rachels 1975; Steinbock 1980). Put differently, any set of circumstances that would morally justify not starting life-sustaining treatment would justify stopping it as well. If this conclusion is correct, then the fact noted above that persons feel more responsible for the patient's death when they stop life support, and are as a result more reluctant to stop than not to start, suggests that many people's natural reactions will here lead them to act in ways that are not morally defensible and that conflict with their own considered moral judgments. Consequently, this may be a place where physicians and others should be especially reflective about their behavior since their unreflective natural reactions may lead them morally astray. It is worth adding that in practice one often gains a reason to stop a treatment once it has been tried that one did not have earlier not to try it. Very often there is considerable uncertainty about how well a patient will do or what progress he or she will make with a particular form of life support such as mechanical respiration. The treatment will then usually be worth trying to see if it in fact has the hoped-for positive effects. When it does not have the hoped-for positive results, and so no longer holds out the reasonable but uncertain prospect of benefits to the patient, there is then a reason to stop the treatment that did not exist earlier not to start it.

Does it matter that many physicians and families are unwilling to stop treatments like respirators in circumstances in which they would be willing not to initiate them, even if there is no significant moral difference between the cases? There are at least two serious bad effects for patients of this reluctance to stop life support. The first and most obvious is patient overtreatment. Life-sustaining treatment will be continued beyond the point at which it is reasonable to believe that the patient either is benefitted by or still wants or would want the treatment. This is wasteful of what is commonly very costly care, but more important it fails to respect the patient's self-determination while often inflicting unnecessary emotional distress on patients, families and others.

The less obvious effect is at least as serious. A very common fear of patients, families, and physicians is that the patient will be "stuck on machines." To avoid this outcome, parties involved in decisionmaking may be reluctant to try life-sustaining treatment when its benefits are highly uncertain. This has the effect of denying life-sustaining treatment to some patients for whom it would have proved to be of genuine and substantial benefit and is indeed a serious harmful consequence of the reluctance to stop life support once it is in place. All parties should be much more willing than is now common to employ time-limited trials of therapies such as ventilator support, in which it is clearly understood that the trial, if unsuccessful, can be terminated, so as to allay this otherwise reasonable fear of patients and families of losing control of treatment and being stuck on machines.

It is worth adding that if the difference between stopping and not starting a life-sustaining treatment is thought to be of moral significance, it is then important how some common cases are to be classified. For example, with any therapy that involves multiple courses of treatment over time, such as dialysis or many medications, a decision to forgo further treatment plausibly might be construed either as stopping the overall course of treatment or as not starting the next course of dialysis or dose of medication. That such cases might be plausibly interpreted either as stopping or not starting should give further pause about resting much moral importance on which is done.

Killing and allowing to die

The distinction between stopping and not starting a life-sustaining treatment corresponds in general to the distinction between acts and omissions leading to death. Ambiguities also abound about whether a decision to forgo a life-sustaining treatment should be classified as an act or an omission. Does the positive decision make what happens an action, or the content of the decision not to start a treatment make it an omission? However these distinctions are more precisely drawn, if there is no moral importance to whether a life-sustaining treatment is stopped or not started, then it would seem to follow that it is not morally significant on this view whether it is an act or omission of the physician that leads to death. This implication has been explicitly accepted in some recent influential commentaries and court decisions (*Barber and Nedjl* 1983; *Conroy* 1985; Wanzer et al. 1984). Yet the distinction between acts and omissions leading to death is also commonly understood to be the basis for the distinction between killing and allowing to die. Some have gone on to accept, or to explicitly argue, that killing is in itself no different morally than allowing to die, though of course many or most actual acts of killing are morally worse, all things considered, than most cases of allowing to die (Glover 1977; Steinbock 1980).

While this view has been increasingly widely accepted by philosophers and bioethicists, many health-care personnel, patients, and their families strongly resist it. For many, that killing is both wrong and also worse than allowing to die is a deeply and powerfully held view. But the positive decision actively to turn off a life-sustaining treatment like a respirator seems to be an action, not an omission, leading to death, and so in this view is considered a killing, not a case of allowing to die. This line of reasoning uncovers a more general concern about whether all stopping of life support is killing and therefore morally wrong. In assessing this question it is important to be clear, first, about the meaning of the claim that killing is *in itself* no different morally than

162

allowing to die. The claim is that the mere fact that one case is an instance of killing, another of allowing to die, does not make one any worse morally than the other, or make one justified or permissible but the other not. This is not to say that any particular instance of killing may not all things considered be morally worse than some instances of allowing to die. It is to say that if the killing is worse it will be because of its other properties such as the motives of the killer, whether the victim consented, and so forth, that differentiate it morally from the particular instance of allowing to die. Secondly, it is important to distinguish whether common instances of stopping life support should be understood as killing or as allowing to die from whether, if they are killings, they are for that reason morally wrong. Most commentators who have argued that stopping life-sustaining treatment is killing have insisted as well that it is not therefore wrong; some killing, including stopping life support, can be morally permissible and justified.

Are standard cases of stopping life-sustaining treatment killing or allowing to die (Brock 1986; Rachels 1975)? A physician who stops a respirator at the voluntary request of a clearly competent patient who is terminally ill and undergoing unrelievable suffering would commonly be understood by all involved as allowing the patient to die, with the patient's underlying disease the cause of death. If done with the consent of the patient, and with the intent of respecting his self-determination while promoting his well-being as he conceives it, it would be held by many to be morally justified. Let us agree that it can be morally justified, but is it allowing to die? Suppose the patient has a greedy nephew who stands to inherit his money and who has become impatient for the old man to die so that he will get the money. Thinking that his uncle is prepared to continue on the respirator indefinitely, he slips into the room, turns off the respirator, and his uncle dies. The nephew is found out, confronted, and replies, "I didn't kill him, I merely allowed him to die; his underlying disease caused his death." Surely this would and should be dismissed as specious nonsense. The nephew deliberately killed his uncle. However, it is plausible that he did exactly what the physician did in the other case. Both acted in a manner that caused the patient's death, expected it to do so, and might have performed the very same bodily movements in doing so. Of course, the physician acts with a different and proper motive, with the patient's consent, and in a professional role in which he is socially and legally authorized to carry out the patient's wishes concerning treatment. These differences in motive, consent and social role make what he does, but not what the nephew does, morally justified. That is not to say, however, that what he does, and whether he kills or allows to die, is any different from the nephew, only that his killing was justified while the nephew's was not. This general line of reasoning, then, accepts that standard cases of stopping life-sustaining treatment can sometimes be correctly understood as killing, but rejects any inference that they therefore must be wrong.

One explanation of why this account is resisted is that many physicians and others use the concept of killing as a normative concept to refer to unjustified actions causing death. In this view, killing may occur in medicine accidentally or negligently, but physicians do not knowingly and deliberately kill their patients. Yet, of course, physicians do stop life support in cases like the above and believe, quite rightly, that they can be justified in doing so. Thus, there is a powerful motive to understand what is done as allowing to die, not as killing. Common though this way of thinking may be, in this analysis it appears confused. It is a mistake to suppose that all killing must be unjustified, either morally or in the law. Killing in self-defense is an example outside of medicine, and stopping life support appears to be one within medicine.

There is another explanation of why standard cases of stopping life support are thought to be allowing to die and not killing. In the case of a terminally ill patient, a lethal disease process is already present. A life-sustaining treatment like a respirator may then be thought of as holding back or blocking the normal progress of the patient's disease. Removing that artificial intervention is then conceived of as standing aside and allowing the patient to die by letting the disease process proceed unimpeded to death. This may be a plausible explanation of why stopping life support is commonly understood to be allowing to die, but if it is to be any more than a metaphorical account, it must at the least explain why the nephew does not also allow to die. It is not clear how this is to be done consistent with the way killing and allowing to die are distinguished over a broad range of cases. It is worth adding that even understanding stopping life support as allowing to die along these lines, it remains possible still to argue that killing is in itself not morally different from allowing to die, though I shall not pursue that argument further here.

Is forgoing life support suicide?

Parallel to the concern that stopping life support is killing and therefore wrong is the concern that any forgoing of life support is suicide or assisted suicide and therefore wrong. Courts in particular try to distinguish forgoing life support from suicide, probably in significant part to insulate physicians, families and other health-care personnel from possible liability under laws prohibiting assisting in a suicide. The recent New Jersey Supreme Court decision *In re Conroy* summarizes well the reasoning of many courts and others:

> Declining life-sustaining medical treatments may not properly be viewed as an attempt to commit suicide. Refusing medical intervention merely allows the disease to take its natural course; if death were eventually to occur, it would be the result, primarily, of the underlying disease, and not the result of a self-inflicted injury. In addition, people who refuse life-sustaining medical treatment may not harbor a specific intent to die, rather, they may fervently wish to live, but to do so free of unwanted medical technology, surgery, or drugs, and without protracted suffering. . . . Recognizing the right of a terminally ill person to reject medical treatment respects that person's intent, not to die, but to suspend medical intervention at a point consonant with the "individual's view respect-

ing a personally preferred manner of concluding life." The difference is between self-infliction or self-destruction and self-determination. (*In re Conroy* 1985)

Although this way of distinguishing forgoing life support from suicide may seem plausible, it is at least problematic in some cases. The judgment of a person who competently decides to commit suicide is essentially that "my expected future life, under the best conditions possible for me, is so bad that I judge it to be worse than no further continued life at all." This seems to be in essence exactly the same judgment that some persons who decide to forgo life-sustaining treatment make. The refusal of life-sustaining treatment is their means of ending their life; they intend to end their life because of its grim prospects. Their death now when they otherwise would not have died *is* self-inflicted, whether they take a lethal poison or disconnect a respirator. There need be, of course, no underlying lethal disease process present when a person commits suicide, whereas there must be when life-sustaining treatment is refused, but that is no reason to think that a person subject to a lethal disease process therefore could not commit suicide.

The Court's reasoning distinguishes some but not all cases of forgoing life-sustaining treatment from suicide, though it should be adequate to protect all instances from falling under legal statutes concerning assisting in suicide. Even if at least some instances of forgoing life-sustaining treatment are suicide that is no reason to conclude that they are morally wrong. The very same reasoning offered earlier in support of a competent patient's moral right to refuse any life-sustaining medical treatment will, of course, apply in these cases in which doing so may be suicide. The patient's self-determination and well-being support the moral permissibility of his or her declining any life-sustaining treatment, including any instance that might reasonably be construed as suicide. Cases of competent decisions to decline life-sustaining treatment that constitute suicide are commonly instances of rational and morally permissible suicide.

Death and dying

Ordinary and extraordinary treatment

A different way of distinguishing some forgoing of life support as impermissible is to hold that while extraordinary treatment can be permissibly forgone, ordinary treatment cannot be. This distinction has had special importance in cases of incompetent patients who are unable to decide about life support for themselves and so must have others decide for them. It is there often held that the surrogate decision-maker can decide to forgo only extraordinary treatment for the patient. Many court decisions too have made reference to extraordinary treatment in endorsing the permissibility of forgoing life support, though often in passing and without any analysis of how ordinary and extraordinary treatment are to be distinguished.

There are two important questions regarding this distinction. First, what is the difference between treatments that the distinction is thought to mark? Second, is that difference of sufficient moral importance to mark a difference between the morally permissible and impermissible? With regard to the first question, it is clear that many different meanings have been intended. Among the differences that the distinction is thought to mark are: treatment that is usual, for example for a given condition, as opposed to unusual; treatment that employs high-technology, artificial means as opposed to relatively simple means; treatment that is highly invasive as opposed to relatively noninvasive; treatment that is very costly as opposed to relatively inexpensive; treatment that is heroic in the sense of a long shot, last ditch attempt to keep the patient alive when other more ordinary means have failed.

Since there are so many interpretations of this distinction in use, confusion about what a particular user intends by it is inevitable unless its meaning is made explicit, as it usually is not. However, for any possible interpretation like those just mentioned, it is important to ask our second question: Why does this difference (for example, whether treatment employs high technology as opposed to relatively simple means) determine whether the treatment is morally justified, or whether forgoing it is morally permissible or impermissible. High-technology respirators or dialysis treatments sometimes promise benefits clearly outweighing any burdens of them so that a competent patient would choose them, while for another patient the life they continue may be of such limited duration and poor quality that the patient will competently choose to forgo them. Why should it be at all morally important to whether the choice is justified that a treatment is common or unusual, employs high technology or is simple, is invasive or noninvasive, and so forth?

Instead, what is relevant is the overall balance of benefits and burdens of the treatment to the patient, according to the competent patient's own assessment, or the incompetent patient's surrogate's judgment of what the patient's assessment would have been if the patient had been competent. Interpretations of the difference between ordinary and extraordinary treatment such as those noted above seem to appeal to differences that are not in themselves morally important, but instead are important only in so far as they affect the benefits and burdens of treatment to the patient.

It can be argued that all of these common and commonsense interpretations of the ordinary/extraordinary difference misunderstand it. The distinction probably originated within Roman Catholic moral theology where extraordinary treatment was understood, roughly, as treatment which was excessively burdensome for the patient (McCormack 1974). When treatment was judged by the patient (or by others acting as surrogates for an incompetent patient) to be excessively burdensome, then it was held that neither patient nor surrogate were obliged to begin or continue it. But of course to determine whether a treatment is *excessively* burdensome it is necessary to weigh its burdens against whatever benefits it promises.

This interpretation of the ordinary/extraordinary difference simply appeals to the patient's assessment of the benefits and burdens of treatment and endorses the patient's right to refuse treatment he or she judges to be excessively burdensome. It thus can constitute no further moral constraint on a patient's right to refuse such treatment that it must be extraordinary and not ordinary. "Extraordinary treatment" is here only the label placed on treatment that has already and independently been assessed as excessively burdensome. The assessment of benefits and burdens, not any independent ordinary/extraordinary difference, is the criterion of whether any treatment is justified. "Ordinary" and "extraordinary" merely label the conclusions of that determination, but play no substantive role in making it. In this understanding, it is misleading to suppose that any list of treatments, some as ordinary, others as extraordinary, is possible which could help determine whether they should be employed with any particular patient. A treatment that is ordinary for one patient in particular circumstances can be extraordinary for another or for that same patient in different circumstances. While the ordinary/extraordinary distinction in this interpretation plays no positive or substantive role in an assessment whether to employ or to forgo a treatment, its use does have one serious bad effect. The many possible and natural understandings of the ordinary/extraordinary distinction mean its use in decisionmaking about life support almost inevitably leads to confusion as different parties understand different things by it. It is for reasons of this sort that many recent commentators have concluded that, except within particular religious traditions where its meaning is clear, the distinction is unhelpful and best avoided in decisionmaking about life support (*In re Conroy* 1985; President's Commission 1983a).

Active voluntary euthanasia

If competent patients are morally entitled to refuse any life-sustaining treatment that they judge to be sufficiently burdensome, so as to make life no longer worth living with that treatment, are they also entitled in similar circumstances to have others, such as their physicians or family members, directly end their lives by administering a lethal injection or a poison? I have deliberately avoided until now using the term "euthanasia" because of its strong emotionally laden connotations, but it is this sort of active killing that is commonly understood as euthanasia. The very same values of patient well-being and self-determination that support a patient's right to refuse any life-sustaining treatment appear to support active voluntary euthanasia or direct killing in some circumstances as well. Does this show that if one accepts that forgoing life support is morally permissible, one must accept that this form of active voluntary euthanasia is permissible as well?

Some have argued that the case for active voluntary euthanasia can be strengthened further still by the argument from mercy (Rachels 1975). Consider the case of a terminally ill and imminently dying patient with a form of cancer that causes him very great and unrelievable suffering. With his competence not in question, the patient implores his physician to end his suffering by giving him a lethal poison. It seems cruelly perverse to hold that if a life-sustaining treatment were in place we should honor the patient's request to remove it and let him die, but that otherwise we cannot intervene and must leave him to suffer in pain until nature runs its course. How could any prohibition of voluntary euthanasia be morally justified if it prevents ending the excruciating suffering of great numbers of dying patients?

This argument from mercy for active voluntary euthanasia would be a powerful one indeed if its factual premises were true, but a crucial premise is false. There are not great numbers of patients undergoing severe suffering that can only be relieved by directly killing them. Modern methods of pain management enable physicians and nurses to control the pain of virtually all such patients without the use of lethal poisons, though often at the cost of so sedating the patient that interaction and communication with others is limited or no longer possible. The vast majority of cases in which such suffering is not relieved are due to wrongful failure to employ effective methods of available pain management, not to a prohibition of active voluntary euthanasia. Even if the unavoidable cost of a prohibition of active voluntary euthanasia is only exceedingly rarely the continuation of severe suffering, however, it does more commonly result in failure to respect the self-determination of patients who want their lives directly ended. This cost should not be lightly borne in a society such as our own that values self-determination highly in both its moral and legal traditions.

What reason is there to bear this cost by prohibiting active voluntary euthanasia? The most influential argument in support of maintaining this prohibition is the so-called "slippery slope" argument (Kamisar 1958). Even granting, so this argument goes, that there are a few cases in which such active killing might be justified, the consequences of socially and legally permitting it overall would nevertheless be bad. If we begin by permitting it in these few cases in which it is justified, we would inevitably end up permitting it in a great many other cases in which it would be wrong. It is the first step on the path to the Nazi policy of killing the old and weak and socially disfavored and must be firmly resisted. Since this path is slippery and steep, we must stay off it altogether.

What is to be made of this argument? If the factual claim is true that any relaxation of the prohibition of active euthanasia must inevitably lead to the Nazis' final solution, then all will agree that the prohibition must be firmly maintained. What is controversial, however, is whether this factual claim is true. We should have some reason or evidence for believing it, not the mere possibility that it might be true. Supporters of active voluntary euthanasia go on to urge that we have no serious reason to believe this claim. For example, we can and do make very clear and firm distinctions between voluntary and involuntary euthanasia; the values supporting the former in no way support the latter, and there is no significant reason to expect that permitting the former must inevitably lead to permitting the latter. This dispute is extremely difficult to resolve just because we possess very little evidence about the truth or falsity of the key factual claim in the argument. No one can reasonably deny the merest possibility that the claim might prove to be true, but on the other hand there is little persuasive evidence that in fact it is true.

There is at least some reasonable worry about possible abuse of an authorization of physicians or others to perform active voluntary euthanasia, even if it is not that it will lead to anything on the order of Nazism. Many frail and debilitated elderly who have become both financial and emotional burdens to others might be made increasingly vulnerable to having their lives wrongly ended by such an authorization. The magnitude of such risks is virtually impossible to estimate and, of course, depends on the exact form such authorizations take and the institutional safeguards built in against abuse. Besides such possible abuse, there are also the potential bad effects on the image of the medical profession, both the public's view and the profession's own view of itself. Public trust in the profession's commitment to fight with the patient against disease and death might be undermined if physicians also become "the angels of death." Physicians themselves might also find the role of euthanasiast in uneasy conflict with their role as medical caregivers to the sick and dying, even to an extent that it undermines their capacity to carry on effectively as caregivers to the dying.

The extent, if any, to which these bad effects would in fact occur from an authorization of active voluntary euthanasia is only speculative. Against these possible bad effects are the very real gains in self-determination and control over the process of dying that such an authorization would yield. Different persons can reasonably reach different conclusions about whether this tradeoff on balance argues in favor of or against permitting this form of euthanasia, but it would appear to be on this basis that the question ought to be decided.

Intended versus merely foreseen consequences

Some commentators give special moral importance to physicians' intentions when they act. They argue that the intentional killing of innocent human beings is morally wrong, while actions from which a person's death is foreseen though not intended may sometimes be morally permissible (Fried 1978). This is the distinction embodied in the Roman Catholic Doctrine of Double Effect, sometimes also characterized as the difference between direct and indirect intention (Frey 1985). This distinction is important in its potential implications for active euthanasia but also for the issue of providing adequate relief of suffering to the dying. The following cases illustrate both implications. It sometimes happens that in the final stages of some terminal cancers adequate levels of medication (usually morphine) to control pain reach levels that risk depressing the patient's respiration and hastening his or her death. In such cases physicians often administer morphine at the patient's request with the intention or goal of relieving the patient's suffering, while foreseeing although not intending the patient's likely earlier death from respiratory depression. Few physicians, on the other hand, would give a lethal injection of strychnine at the patient's request to end the patient's suffering if morphine were unavailable or unavailing. Apart from what is legally permitted, is there an important moral difference between the two cases?

Many have thought that the difference lies in the physician's intentions. No fully adequate analysis exists of the concept of intention. Nevertheless, it does seem that the patient's earlier death is intended only in the strychnine, not in the morphine case. In each case, however, the physician's aim is to respond to the patient's request to end his suffering. The difference appears to be that in the strychnine case the means used to do so is to kill the patient; only through his death is the suffering ended. In the morphine case the administration of morphine is the means to the end of relieving the suffering while the earlier death is merely a foreseen side effect. The end sought is the same in each case, and the difference is that the death is the means to the end in one case and the foreseen consequence following the end in the other.

Can this difference be of sufficient moral importance to make the one morally permissible and the other prohibited? Many have argued that it cannot (Bennett 1981). In each case the physician's end or motive of relieving suffering at the request of the patient is the same. In each case it is causally impossible to end the patient's suffering without acting in a way that will cause his death. In each case both the patient and physician are prepared to end the suffering even at the cost of the patient's earlier death. The relief of suffering is judged to be of sufficient importance to justify acting in a way that leads to death. These seem to be the essential value judgments involved and they do not differ in the two cases. The difference in intention seems to be one of temporal structure – in one, the death precedes the end of relief of suffering, while in the other it temporally follows the end. It is hard to see why this difference in temporal structure should have much, or any, moral importance. It is tempting to reply that in the morphine case one would have given the morphine even if respiratory depression and death would not have followed, while the point of the strychnine was to cause death. However, if somehow strychnine would have relieved the suffering without causing death one would have still given it as well. In each case, in the circumstances that existed it was necessary to act in a way that the physician knew would lead to the patient's death in order to relieve his suffering.

There is a difference in the two cases in the certainty with which the earlier death will occur. It may never be certain that the dosage level of morphine is sufficient to cause death, and none would deny that this is a morally significant difference in the two cases. However, in some instances this difference in probability may be extremely small and so not support a great moral difference between the two cases. In any event, it is a difference in the risk of a bad outcome and not of intentions. Critics of this foreseen/intended distinction have argued that physicians are reasonably held equally morally responsible for all the foreseen consequences of their actions, whether or not intended, since all such consequences are under their control. In this perspective, in both cases it is a matter of weighing the relative benefits and burdens to the patient of relieving his suffering and shortening his life.

If the patient judges that relief of suffering is paramount, the physician would be morally justified in acting in either the morphine or strychnine case. For reasons of public policy (see Chapter 8), it might be wise, nevertheless, not legally to authorize the performance of active voluntary euthanasia as in the strychnine case. It is important to emphasize that this would not be because the two cases are in themselves significantly different morally, but because of public policy concerns about the one and not the other. It is important to emphasize as well that the general right of the patient to decide about treatment does include the right to have adequate pain medication even if that may shorten his life. The relief of suffering is a long-standing, central, and fully legitimate aim of medicine.

SOME CASES OF SPECIAL POLICY CONCERN

Many of the issues addressed in this chapter are of very great public and policy concern. I have sought above to cover the principal questions involved in the development of an overall ethical framework for decisions about life support. Of course, no such framework can be applied in any mechanical fashion to yield conclusions in particular cases. A framework is only that, and it must be applied with sensitivity and understanding to the unique features and details of any actual case. Nevertheless, the ethical framework should be applicable across the broad class of life-support cases. There are, however, two issues of special current concern that raise some questions not yet addressed: seriously ill newborns and life-sustaining nutrition and hydration.

Seriously ill newborns

Several cases of forgoing life support for seriously ill newborns have become front page news in recent years. Public and government attention was focused in this area several years ago in the so-called "Baby Doe" case in Bloomington, Indiana, when an infant who suffered from Down's syndrome was allowed to die after its parents refused to permit surgery to repair its esophagus so that it could take nourishment. The federal government promulgated its "Baby Doe Rules," which went through extended negotiations and court challenges. These rules essentially require that all medically indicated treatment be provided to an infant unless the infant is irreversibly comatose, the treatment would merely prolong the dying of the infant and would be futile in terms of its survival, or the administration of the treatment would itself be virtually futile and inhumane. It is not yet clear how these proposed regulations will be interpreted or what their impact will be, but it is clear that their intent, both symbolic and for practice, was both to exclude the use of considerations about an infant's expected quality of life from decisions about its treatment and to seriously limit parents' and physicians' discretion in decisions to forgo life support for such infants.

While the fundamental ethical framework developed above for life-support decisions generally applies to newborns, nevertheless, there are a few issues that are especially prominent with newborns (Singer and Kuhse 1986; Weir 1984). One is already implicitly noted in the proposed Baby Doe regulations. What role, if any, should the infant's expected quality of life play in decisions about treatment? As discussed earlier, a narrowly constrained role for quality-of-life considerations is inevitable if competent patients, or incompetent patients' surrogates, are to be free to decide whether a life-sustaining treatment and the life that it makes possible are on balance a benefit or excessively burdensome. Using this standard, the infant's prospects in relatively few cases will be so poor that it is reasonable to hold that continued life is clearly not in its interests. The clearest cases are probably when its life will be filled with substantial and unrelievable suffering and when the infant has suffered such severe brain damage as to preclude any significant social or environmental interaction. Other cases of very severe disabilities are more controversial and problematic, in part because of the wide variation in the weight adult patients give to such considerations and the fact that infants do not yet have preferences or values of their own.

A second issue is the relevance, if any, of the effects on others such as the parents of continued life support for the newborn. As a general matter our society rejects the involuntary sacrifice of one person's life for the benefit of others. The very high value we give to the protection of human life suggests that life-sustaining treatment for a seriously ill newborn should rarely if ever be forgone because of the burdens its continuing existence would place on others. Moral rights not to be killed are commonly understood to protect an individual from being killed, whatever the effects on others of the individual continuing to live. Nevertheless, it would be callously insensitive to deny the often overwhelming long-term burdens placed on the parents of these newborns, particularly in light of the often inadequate support services available to them. This issue reflects a fundamental difference between utilitarian moral conceptions, which weigh all the effects of a decision about treatment including the effects on others besides the patient, and moral rights conceptions, which exclude effects on others from consideration when a person's right to life is in question. As a practical matter, in most cases in which the parents judge the infant to be an excessive burden on them they have the alternative of giving it up for adoption.

A third issue concerns the moral status of infants. Should they be given the very same moral (and legal) protections as adults, particularly in light of common views on the moral permissibility of aborting fetuses? Birth seems a problematic point at which to draw a great moral difference in the moral permissibility of killing. Some commentators consider newborns closer morally to unborn fetuses than adults, or as somewhere between the two (Tooley 1983). Unborn fetuses, however, in this view are often considered replaceable and permissibly killed – for example, when found by amniocentesis to be defective – in order to try again for a normal pregnancy. This general issue concerns whether infanticide might be morally permissible in at least some circumstances in which killing an adult person would not be.

Finally, there is the policy issue of what review mechanisms, if any, are needed for parents' and physicians' decisions to forgo a life-sustaining treatment for a newborn. There are at least two important reasons to believe that conflicts of interest between seriously ill newborns and their parents may be more common than in most other cases of surrogate decisionmaking for incompetent patients. First, the strong bonding that exists between parent and older child has not yet occurred with newborns, which can result in a weaker commitment of the parent to the infant's well-being. Second, the often enormous burdens of caring for the infant if life-sustaining treatment is continued will fall largely on the parents. Thus, serious conflicts of interest may be sufficiently likely in decisions to forgo life-sustaining treatment of newborns to warrant some form of regular review, for example, by hospital ethics committees, so-called infant care review committees, or even the courts.

These are some of the issues of special importance in life-support decisions for newborns, but it bears emphasis that the fundamental ethical issues are not different from those with adults.

Life-sustaining nutrition and hydration

The moral permissibility of a competent patient, or an incompetent patient's surrogate, deciding to forgo treatments such as respirators or kidney dialysis has become fairly widely accepted by the public and health-care professionals, as well as in recent court decisions. More recently, concern has focused on the provisions of artificial nutrition and hydration through the use of nasogastric tubes, intravenous lines, surgically inserted tubes, and so forth. Must nutrition and hydration, or food and water, always be continued or may it too permissibly be forgone (Lynn 1986)?

If nutrition and hydration are considered part of treatment, the same ethical framework discussed above for other kinds

of life support can be applied to them. That framework calls
for an assessment of the benefits and burdens to the patient
of continuing nutrition and hydration as opposed to discontinuing it. For nearly all patients that assessment will favor
continuing them, but in a few cases it need not do so (Lynn
and Childress 1983). The process of providing nutrition and
hydration can itself involve substantial ineliminable burdens
for patients – for example, when patients' disease states make
taking in nutrition the cause of significant discomfort or when
patients' dementia and resultant confusion requires physically restraining them to prevent their removing feeding
tubes. In a very few other cases the quality of the life continued by nutrition and hydration may be substantially and
unalterably burdensome to patients. Against possible burdens of continuing nutrition and hydration must be weighed
the benefits and burdens of discontinuing it. For most patients, discontinuing nutrition and hydration would result in
substantial subjective distress and likely not be in their interests even if death might otherwise be welcomed by them.
Once again, however, for a few patients in the final stages
of certain terminal diseases or in a persistent vegetative state
in which all conscious experience is irretrievably lost, withholding nutrition and hydration and their resulting death by
starvation or dehydration does not result in significant suffering. It bears repeating that the assessment of the benefits
and burdens to a particular patient of continuing life-
sustaining nutrition and hydration will in the vast majority
of cases favor doing so, but there can be no assurance that
this must always be so.

This conclusion has been challenged on several grounds:
(1) human life is of infinite value and so deliberately shortening it can never be a benefit to the victim; (2) a patient's
choice to forgo food and water is suicide and therefore wrong;
and (3) life is a gift of God that we are not at liberty to destroy.
I shall not pursue these objections in any detail here. The
first two are really general challenges to forgoing any life
support, not nutrition and hydration in particular, and so
understood have already been addressed. The third appeals

to religious views which could quite properly guide the choices of those individuals who share the particular religious faith, but should not form the basis for public policy in a pluralistic society.

A further concern about permitting the forgoing of nutrition and hydration centers on its symbolic effect together with worries about abuse (Siegler and Weisbard 1985). The provision of food and water is one of the very first acts of concern and support each person receives as he or she enters the world. Throughout life, feeding the hungry is properly invested with great moral importance, and starvation is associated with suffering and strong moral repugnance. This deep concern with not permitting suffering or death from starvation in general serves us well and, in this view, we should be extremely reluctant to do anything that might weaken it. This is especially so since many of those who might be endangered by any weakening of the requirement to provide food and water are debilitated and vulnerable, unable to protect themselves and to assert their own interests. Permitting the withholding of food and water risks the deliberate killing of the vulnerable and burdensome on morally unacceptable grounds such as the economic costs of sustaining their lives. Moreover, many see any withholding of food and water as only a very small step from active voluntary or even involuntary euthanasia and oppose the former for fear that it will inevitably lead to the latter.

Like all "slippery slope" arguments based on a worry about potential abuse of a specific authorization, reasonable persons may disagree on several counts. Is it possible to discriminate clearly between cases in which withholding food and water is morally justified from those in which it is not? What procedures for making such decisions would provide the most effective safeguards against abuse? Given such procedures, how likely are abuses and are they sufficiently serious to outweigh the benefits of permitting withholding food and water that is excessively burdensome to the patient? Courts that have addressed the issue of withholding nutrition and hydration have been sensitive to the need for procedures with strong safeguards against abuse, but have nonetheless generally held that artificial measures for provision of nutrition and hydration are not in principle different than other treatments such as respirators that artificially provide oxygen to the patient (*Barber and Nedjl* 1983; *In re Conroy* 1985). All fall under a patient's general moral and legal right to decide about and to refuse any medical treatment.

CONCLUSION

I have sought in this chapter to address many of the most difficult and troubling ethical issues arising in decisions about life-sustaining treatment. Many of those issues are complex and no widespread agreement yet exists on them. I shall let the discussion of those issues in the body of the chapter stand on its own and not attempt to summarize it here. There is, however, a central core of widespread agreement that is worth repeating and underlining. A competent patient, or an incompetent patient's surrogate, is ethically entitled to assess the benefits and burdens of any proffered treatment according to the patient's own aims and values and to accept or reject the treatment. The quite broad agreement on this seemingly simple principle should in no way be taken to imply, however, that the decisions themselves in concrete cases are simple or uncontroversial for those involved in them. There is much truth in the view that the particular circumstances and details of each case make it unique. No simple principles can be mechanically applied in a way that makes for easy choices. Decisions about life and death are inevitably and quite properly difficult and troubling and require sensitive, thoughtful and wise judgment from all involved.

REFERENCES

Barber and Nedjl v. Superior Court. 195 Cal. Rptr. 484 (Cal. App. 2 Dist.) 1983.

Bayer, R., et al. "The Care of the Terminally Ill: Morality and Economics." *N Eng J Med* 309 (15 Dec. 1983): 1490–1494.

Bennett, J. "Morality and Consequences." In *The Tanner Lectures in Human Values II*, edited by S. M. McMurrin. Salt Lake City: University of Utah Press, 1981.

Brock, D. W. "Forgoing Food and Water: Is It Killing?" In *Forgoing Life-Sustaining Food and Water*, edited by J. Lynn. Bloomington: Indiana University Press, 1986.

Buchanan, A., and Brock, D. "Deciding for Others." *The Milbank Quarterly* 64, Suppl. 2, (1986): 17–94.

Capron, A. M. "The Authority of Others to Decide About Biomedical Interventions with Incompetents." In *Who Speaks for the Child?* edited by W. Gaylin and R. Macklin. New York: Plenum Press, 1982.

Capron, A. M., and Kass, L. "A Statutory Definition of the Standards for Determining Human Death: An Appraisal and a Proposal." *U Penn Law Rev* 87 (1972): 102–104.

In re Conroy. 486 A.2d 1209 (N.J. 1985).

Drane, J. "The Many Faces of Competency." *Hastings Cent Rep* 15, no. 2 (April 1985): 17–21.

Dworkin, G. *The Theory and Practice of Autonomy*. Cambridge, Mass.: Cambridge University Press, 1988.

Fost, N., and Cranford, R. "Hospital Ethics Committees: Administrative Aspects." *JAMA* 253, no. 18 (10 May 1985): 2687–2692.

Frey, R. "Some Aspects to the Doctrine of Double Effect." *Can J Philos* (1985): 259–283.

Fried, C. *Right and Wrong*. Cambridge, Mass.: Harvard University Press, 1978.

Glover, J. *Causing Death and Saving Lives*. New York: Penguin Books, 1977.

Green, M., and Wikler, D. "Brain Death and Personal Identity." *Philosophy and Public Affairs* 9 (1980): 105–133.

Kamisar, Y. "Some Non-Religious Views Against Proposed Mercy Killing Legislation." *Minnesota Law Review* 42 (1958).

Kass, L. *Toward a More Natural Science: Biology and Human Affairs*. New York: Free Press, 1985.

Katz, J. *The Silent World of Doctor and Patient*. New York: Free Press, 1984.

Lynn, J., ed. *Forgoing Life-Sustaining Food and Water*. Bloomington: University of Indiana Press, 1986.

Lynn, J., and Childress, J. "Must Patients Always Be Given Food and Water?" *Hastings Cent Rep* 13 (1983): 17–21.

McCormack, R. "To Save or Let Die: The Dilemma of Modern Medicine." *JAMA* 229 (1974): 172–176.

President's Commission for Ethical Problems in Medicine. *Defining Death.* Washington, D.C.: Government Printing Office, 1981.

———. *Making Health Care Decisions.* Washington, D.C.: Government Printing Office, 1982.

———. *Deciding to Forego Life-Sustaining Treatment.* Washington, D.C.: Government Printing Office, 1983a.

———. *Securing Access to Health Care.* Washington, D.C.: Government Printing Office, 1983b.

In re Quinlan. 70 N.J. 10, 355 A. 2d 647 (1976).

Rachels, J. "Active and Passive Euthanasia." *N Engl J Med* 292 (9 Jan. 1975): 78–80.

Rosner, F. "Hospital Medical Ethics Committees: A Review of Their Development." *JAMA* 253 (1985): 2693–2697.

Sider, R., and Clements, C. "The New Medical Ethics: A Second Opinion." *Arch Intern Med* 145, no. 12 (1985): 2169–2171.

Siegler, M. "Searching for Moral Certainty in Medicine: A Proposal for a New Model of the Doctor-Patient Encounter." *Bull NY Acad Med* 57 (1981).

Siegler, M., and Weisbard, A. "Against the Emerging Stream." *Arch Intern Med* 145 (1985): 129–131.

Singer, P., and Kuhse, H. *Should This Baby Live?* New York: Oxford University Press, 1986.

Steinbock, B., ed. *Killing and Letting Die.* Englewood Cliffs, N.J.: Prentice-Hall, 1980.

Tooley, M. *Abortion and Infanticide.* Oxford, England: Oxford University Press, 1983.

Veatch, R. "The Whole-Brain Oriented Concept of Death: An Out-Moded Philosophical Formulation." *J Thanatology* 13 (1975).

Wanzer, S., et al. "The Physicians' Responsibility Toward Hopelessly Ill Patients." *N Engl J Med* 310, no. 15 (12 April 1984): 955–959.

Weir, R. *Selective Non-Treatment of Handicapped Newborns.* New York: Oxford University Press, 1984.

Chapter 7

Forgoing life-sustaining food and water: Is it killing?

The moral permissibility of patients forgoing life-sustaining medical treatment has come to be widely accepted. The issue of forgoing life-sustaining food and water, however, has only very recently gained attention in public policy discussions. One source of resistance to extending this acceptance of a general right to forgo life-sustaining treatment to the case of food and water has explicitly philosophical origins: for a physician to withhold food and water might seem to be not merely to allow the patient to die, but to kill the patient, and therefore wrong. A closely related moral worry is that for physicians to withhold food and water would be to make them the direct cause of their patients' deaths, which also would be wrong. And finally, many worry that providing food and water is ordinary care, not extraordinary or "heroic," and so must be obligatory.

In each case, a distinction is drawn – between killing and allowing to die, causing or not causing death, and withholding ordinary or extraordinary care – and in each case it is claimed that the former, though not the latter, is morally forbidden. I consider appeal to the intrinsic moral importance of these distinctions to be confused, both in general and as applied to food and water. In the hope of reducing the impact of these moral confusions in the policy debate about forgoing food and water, I will address here both the general meaning and the putative moral importance of these distinctions, as well as their specific application to the case of food and water. The upshot of my argument will be that forgoing food and water does not fall under any special moral prohibitions that would make it in itself morally different than the forgoing of other life-sustaining medical care. I believe that a competent patient has the moral right to forgo any life-sustaining treatment, including food and water. If the patient is incompetent, as is usually the case when foregoing food and water is seriously at issue, the surrogate's decision should reflect what the patient would have wanted if competent, or, in the absence of knowledge of the patient's preferences, reflect an assessment of the benefits and burdens to the patient.

I. KILLING AND ALLOWING TO DIE

Is it killing to forgo food and water? And why is that thought to be important? Since forgoing food and water is obviously behavior leading to death and is known to be such at the time it is done, why is it thought important to ask whether it is killing?

There is a common view, among physicians and much of the general public, that physicians can allow patients to die by stopping life-sustaining treatment, but they cannot kill patients. In this view, killing is wrong, and it occurs in the medical context only as a result of accident or negligence. This is to use the concept of killing normatively, to capture in the category of killings only wrongful actions leading to death. Physicians do, however, stop life-supporting treatments frequently in medical contexts, and rightly believe that they are generally justified in doing so. If they believe that killing is wrong, and occurs in medical contexts only as a result of accident or negligence, then they have a strong motive for interpreting what they do when they stop life support as allowing the patient to die, but not as killing.

I think this interpretation of all stopping of life support as allowing to die is problematic and leads in turn to worries about whether stopping life support is morally justified. Let me, therefore, address directly what the difference is between killing and allowing to die. I will offer two interpretations of that difference: on the first, most stopping of food and water – and of other life-sustaining treatments – will turn out to be killing; on the second, allowing to die.

In the first interpretation, the distinction between killing and allowing to die is the distinction between acts and omissions leading to or resulting in death. When I kill someone, I act in a way that causes that person to die when they would not otherwise have died in that way and at that time. When I allow someone to die, I omit to act in a way that I could have acted and that would have prevented that person dying then; that is, I have, and know that I have, both the ability and the opportunity to act to prevent the death, but fail to do so.

Now suppose that the difference between killing and allowing to die is understood as I have just described it. Consider the case of ceasing respirator support: the respirator is turned off and the patient dies. Suppose this is done by a physician who believes it to be justified and who does it with the patient's consent. The patient may have asked to be allowed to die and may understand what is done as allowing him or her to die. But according to this first interpretation of the difference between killing and allowing to die, what the physician has done is to kill the patient, since in turning off the respirator he or she has acted in a way that causes the patient to die in that way and at that time. Now, is that mistaken? Is it mistaken to say that, by turning off the respirator, the physician killed the patient? Physicians, at least, do not commonly understand what they do when they turn off burdensome respirators with patients' consent as killing the patients.

To help see that this might be correctly construed as killing, consider the case of a nephew who, impatient for his uncle to die so that he can inherit the uncle's money, turns off the uncle's respirator. In this case, I think most would understand the nephew to have killed his uncle. We would take it as a piece of sophistry if the nephew defended himself by replying, "No, I merely allowed my uncle to die," or, "I didn't kill my uncle; it was the underlying disease requiring the use of a respirator that killed him."

The difference in the two cases, in my view, is not in *what* the physician as opposed to the nephew does. It is not a difference between killing and allowing to die. The difference is the presence of other morally important factors, most obviously the difference in motivations of the nephew and the physician, as well as whether the patient consented to the respirator being stopped. The difference is not that one and not the other kills, but that one and not the other kills justifiably.

Thus, if the difference between killing and allowing to die is interpreted in terms of the difference between acts and omissions resulting in death, it would seem that when one stops a life-sustaining treatment like a respirator expecting that a patient will die as a result, one thereby kills the patient. My examples also suggest that in some cases one does so with justification, in others not. But is this difference between killing and allowing to die of moral importance? Does the mere difference that one kills in one case, allows to die in the other, *in itself* make the one any more or less morally justified, or wrong, than the other?

Consider the following case:

A patient is dying from terminal cancer, undergoing great suffering that cannot be relieved without so sedating him that he is unable to relate in any way to others. This patient prepared an advance directive in an early stage of his disease indicating that in circumstances like this he wished to have his life ended either by direct means or by withdrawing life-sustaining treatment. In a recent lucid moment he reaffirmed the directive. The attending physician and the patient's family are in agreement that the patient's desire to die ought now to be granted.

Now consider two alternative conclusions to this case. In the first, while the patient is deeply sedated and not conscious, his wife places a pillow over his face and he dies peacefully from asphyxiation. In the second, while the patient is deeply sedated and not conscious, he develops severe respiratory difficulty requiring emergency professional intervention and use of a respirator if he is to live. His wife knows this, is present, and decides not to call the professional staff and thereby not to employ the respirator. Her goal is to allow him to die. The patient dies peacefully.

Suppose we still understand the difference between killing and allowing to die in terms of acts and omissions resulting in death. In the first outcome, the patient is killed by his spouse, while in the second outcome, the spouse decides not to treat or not to seek treatment and thereby allows him to die. Is there any reason why what the wife does in the first instance is morally (as opposed to legally) worse or different from what she does in the second instance? If not, then what one does if one kills is *in itself* no worse morally than if one allows to die. Of course a particular act of killing may be worse, when other features of the action are considered, than a particular allowing to die. But so also some allowings to die may be morally worse, all things considered, than some killings. And if the mere fact that in one case a person kills, in another allows to die, is not itself a difference of moral significance, then it is also not morally important whether stopping a respirator is interpreted as killing or as allowing to die. Some instances of killing and of allowing to die will be morally wrong; some of each will be morally justified. That what one does is to kill or to allow to die will usually count as a moral reason against doing it; which it is considered to be, however, will not make what one does more or less justified or wrong. Other factors, such as whether the patient consents, and what sort of life could otherwise be offered, will morally differentiate particular killings and allowings to die.[1]

The act/omission difference is the first interpretation of how killing differs from allowing to die, and on that interpretation, as I have briefly argued, killing is not morally worse in itself than allowing to die. There remains, nevertheless, strong resistance to accepting that physicians kill when they stop life-sustaining treatment, as follows on the act/omission interpretation. I have already noted one reason for this resistance: if killing is understood as a normative category, referring only to wrongful actions leading to death, then physicians quite justifiably do not want to understand what they do as killing and therefore wrongful when they stop life-sustaining treatment. They are quite correct that it need not be morally wrong.

There is a second reason for resistance to understanding stopping life-support treatment as killing, and that is a different interpretation of the difference between killing and allowing to die. Very loosely, the distinction is this. If you kill someone, what you do is to initiate a deadly causal process that leads to the person's death. If you allow someone to die, you allow a deadly causal process which you did not initiate to proceed to its result of a person's death.[2] One way to allow to die is simply to omit to act in a way that would have prevented the death. That is an allowing to die on the act/omission understanding of the killing/allowing to die distinction. But another way to allow to die is to *act* in a way that allows some deadly causal process, which at present is halted, to follow out its course so as to result in death.

Let me say a little bit more about this second way of allowing to die, since it is essential to the common understanding of stopping life support as allowing to die. Why does a physician who stops a respirator allow a patient to die in this account of allowing to die? There is a life-sustaining process in place – the respirator. There is a deadly disease process present which requires the use of a life-sustaining respirator, and which is, in effect, being held in abeyance by the use of the respirator. By the positive action of turning off the respirator, the physician then stops that life-sustaining process and allows the deadly disease process to proceed; the physician thereby allows the patient, as it is often put, to die of the underlying disease. Now, in this account of the killing/allowing to die distinction, the physician does allow the patient to die when there is an independent underlying disease process which is being held in abeyance and the physician takes positive action to remove the system which is holding that underlying disease process in abeyance.

But what about our earlier greedy nephew? In this account of allowing to die, doesn't he too, like the physician, allow to die? Doesn't he do the same as the physician does, though with different motives? He too merely allows the disease process to proceed to conclusion. Proponents of this second interpretation of the kill/allow to die distinction will be hard pressed to avoid accepting that the nephew allows to die, though they can add that he does so unjustifiably. Some will take this implication for the greedy nephew case to show that this second interpretation of the kill/allow to die difference is not preferable to the first act/omission interpretation after all.

There is not space here to argue that the difference between killing and allowing to die on this second interpretation of the distinction also lacks moral importance. I believe that is so, and that pairs of examples similar to the one cited above for the act/omission interpretation can be constructed to show that.

With these two interpretations of the killing/allowing to die distinction now before us, we are finally in a position to consider whether forgoing food and water is killing. Suppose that a decision has been made not to feed a patient and that the choice has been made according to sound procedures and in circumstances in which all concerned agree this course is morally justified. All further feedings are omitted and the patient dies of dehydration or malnutrition. But then the question is raised whether what was done killed the patient or allowed him to die. Since we have already noted more than one interpretation of that difference, we know the question is more complex than it might seem. On the act/omission interpretation of the difference, since we have said that feedings were omitted, it would seem that the patient was allowed to die. On the other hand, if, for example, IV's were in place and a physician had to take positive action to remove them, it would seem we then have an action leading to death, and so a killing. This shows the difficulty in practice of applying the kill/allow to die distinction in its act/omission interpretation. But it also displays the lack of any moral significance in this difference between killing and allowing to die. Suppose that just before the physician enters the patient's room to remove the IV, it falls out on its own, and upon seeing this the physician then deliberately omits to reinsert it so that the patient may die. Could there really be any moral importance attached to whether the physician removes the IV or fails to reinsert it in this case?

To think of omitting to feed the patient as allowing him to die is in keeping with common understandings of related cases. Suppose you consciously omit to send food to persons whom you know are starving in a famine in some distant land. The famine victims die. No one would say that you killed them by not sending food, but rather that you allowed them to die (whether or not you were morally wrong to have done so). Why do some persons, nevertheless, want to use the more active concept of killing and to say that one kills by denying food and water in the medical context? One reason lies in the confusion of thinking that one more strongly morally condemns what is done by calling it killing. If I am correct that killing is in itself no worse than allowing to die, but instead other factors morally differentiate actual killings and allowings to die, then withholding feeding is no worse morally for being killing instead of allowing to die.

There is a second reason why some persons might understand denying food and water to a patient as killing, and it reveals a further complexity in our moral thinking about this issue. I think the clue here is found in the naturalness of speaking of *denying* food and water to the patient. To speak in terms of denying food and water is implicitly to assume that giving food and water in a medical context of patient care is the expected course of events. It is seen as statistically expected, as what is standardly done, or, perhaps at least as important, as what is morally required. It is assumed that the statistically and morally expected course of events is that patients will be given the basic care of food and water, that this is part of the health care professional's moral obligation to care for the patient. If so, then if some patient is not to receive food and water, someone must actively intervene to stop this normal course of events. The decision to omit to give food and water is seen as an active intervention in the normal caring process, making death occur when it would not otherwise have occurred. And so this positive decision not to feed, even if resulting in an omission to feed, is seen as killing. It is important to understand, however, that this line of reasoning gains its plausibility from standard cases in which the patient is able to eat and drink in normal ways and is clearly benefited by being provided with food and fluids. In those cases, providing food and fluids is both statistically and morally expected. In other cases, however, providing food and fluids requires sophisticated medical procedures, for example, total parenteral nutrition, and may not be of benefit to the patient; then, the assumption that feeding is either statistically or morally expected may be unwarranted. But without that assumption, omitting to feed should not be understood as an active denial of nutrition and therefore as killing.[3]

Some commentators have cited a further reason why stopping feeding, unlike stopping most other forms of life-sustaining treatment, is not allowing to die on the second interpretation I offered above of the killing/allowing to die difference.[4] In that interpretation, we allow to die when, by act or omission, we allow a deadly disease process that is being held in abeyance by a life-sustaining treatment to proceed to death. For example, when a respirator is withdrawn, the patient is allowed to die of the underlying disease requiring use of a respirator. But when food and water are withheld, it is said, we introduce a *new* process – that of dehydration or malnutrition – which will result in death. The patient dies from this new causal process we introduce, not from any already present fatal disease process that was being held in abeyance by a life-sustaining treatment. We thereby kill the patient. However, this line of reasoning appears to be mistaken.

Whenever a disease process attacks a patient's normal ability to eat and drink, and artificial means of providing nutrition are required, then feeding by artificial means can be seen as a form of life-sustaining treatment. Forgoing feeding when IV's, nasogastric tubes, and so forth, are required is then to forgo employment of a life-sustaining treatment – artificial provision of nutrition – and to allow the patient to die from a disease that has impaired his normal ability to eat and drink. It would seem to be only when the patient's normal human ability to take in nutrition is unimpaired, and a decision is then made not to sustain life and so to stop feeding, that a new fatal process is introduced as opposed to withdrawing a life-sustaining treatment and letting the disease process proceed to death. The vast majority of cases of forgoing food and fluids in the medical context are of the former sort, and so constitute allowing to die, not killing on the second interpretation of that distinction.

Let me summarize my discussion up to here, since the analysis has become rather complicated. (In my defense, I believe that this complexity is not merely the result of philosophers making simple matters appear complex, but arises instead from existing complexities and confusions in common, nonphilosophical thinking on these issues.) Does one kill, or does one allow to die, by forgoing food and water? On the act/omission interpretation of the difference between killing and allowing to die, one kills if the forgoing involves a positive action to stop feeding, allows to die if one only omits to feed. In cases in which the expected course of events is feeding, any forgoing of feeding even if it results in an omission to feed, can be seen as actively intervening to change that course of events, and so as killing; however, in many actual forgoings of food and fluids it cannot be assumed that feeding is either the statistically or the morally expected course of events. I have also argued that the difference between killing and allowing to die on its act/omission interpretation is of no moral importance, and so whether forgoing food and water is in any instance morally justified will turn entirely on other matters than whether it is killing as opposed to allowing to die.

In the second interpretation of the kill/allow to die difference, stopping a life-support system like a respirator is understood as allowing the patient to die from the underlying disease process held in abeyance by the respirator. But to forgo food and water is generally to stop the life-sustaining artificial provision of nutrition and hydration, and so to allow the patient to die of the underlying disease process that has impaired his normal human ability to eat and drink; thus, it too is usually to allow to die. However, I have also suggested, though not argued, that, on this second interpretation as well, whether any instance of forgoing food and water is morally justified turns entirely on other questions than whether it is killing as opposed to allowing to die.

II. CAUSE OF DEATH

I turn now to a related confusion in much discussion of forgoing life-sustaining treatment that concerns a different aspect of the issue about the cause of death than that discussed above. It is related because many persons hold that to stop a life-support process such as feeding is to kill, because one directly causes the patient's death; whereas, when one allows to die, it is the underlying disease process, not the physician, that causes the patient's death.

Questions of causality are exceedingly difficult and complex, but I will try at least briefly to illuminate two important confusions about causing death prominent in discussions about life-sustaining treatment generally, and relevant to forgoing food and water in particular. Consider what can be called a "but for" sense of causality: but for this, that wouldn't have occurred, but for Jones poisoning the food Smith later ate, Smith would not have died. Consider our greedy nephew again. But for his turning off the respirator, his uncle wouldn't have died: the nephew, therefore, causes the uncle's death. As I suggested above, it would be absurd for the nephew to assert, "No, not I but the underlying disease caused my uncle's death." Now consider the physician who seemingly does the same thing and stops the patient's respirator: he too satisfies the "but for" condition for the patient's death. But for the physician's stopping the respirator, the patient would not have died; the physician causes the patient's death. Finally, consider stopping feeding. But for that, the patient would not have died; the physician's withholding food and water causes the patient's death.

What I have called a "but for" sense of causality is *not* by itself an adequate account of ordinary attributions of causality. There, one who kills another (e.g., by giving poisoned food) is considered the cause of the other's death, whereas one who allows another to die (e.g., by not providing food to a famine victim) is not generally said to have caused the other's death. The point is that the broader "but for" sense of causality seems to provide the necessary control over the outcomes of what a person does ("does" interpreted broadly to include both acts and omissions) to allow ascription of at least prima facie moral responsibility to the person for the outcomes. And the "but for" causality condition for the death is satisfied both when one kills and when one allows to die, and equally by the greedy nephew, the physician stopping the respirator, and the physician stopping feeding. Why then should we mark any moral difference in these cases because of supposed causal differences when the "but for" sense of causality is equally satisfied in all of them?

I will consider two responses to my claim that there is no morally important causal difference in these cases, each of which, I think, brings out something interesting about causing death, and about why causal talk is often confusing in this area. The first response is associated with talk of "merely prolonging the dying process," and goes like this: If a patient is terminally ill, then a physician who stops the patient's life-sustaining treatment neither kills nor allows him to die. The physician doesn't kill him; rather, the fatal disease does (more on this shortly). Nor does the physician allow him to die, because to allow someone to die it has to be possible for you to save the person; if the patient is terminally ill, you couldn't save him. So on this response, there's nothing the physician does that causes death, nothing that but for doing it the patient would live.

There is more than one confusion in this response. Surely one *can* either kill or not save a terminally ill person. When one either kills or allows to die, what one does causes death in the "but for" sense of causality. This can only mean that one causes a patient to die *at a particular time* (and, to avoid certain complications of causal overdetermination, in a particular way). No one ever prevents someone's death completely, without any qualifications to its being at a particular time (or in a particular way), because we can't make anyone immortal. What we do is to make or allow death to occur earlier than it otherwise would have done, and in that sense what we do does causally affect the person's death, whether or not the person was terminally ill.

It is worth nothing that the claim that a physician is never a "but for" cause of the patient's death would not help in some important cases of stopping food and water even if it was itself sound. I have in mind, especially, permanently unconscious patients who are usually not terminally ill in any plausible sense of "terminally ill." These patients can, and often do, survive for many years in that state. This ought to give some pause to the many commentators who seem prepared to agree that stopping food and water, or for that matter stopping other life-sustaining treatment, can only be justified if the patient is terminally ill.[5] If one believes, as I do, that stopping life-sustaining treatment, including food and water, can be morally permissible with permanently unconscious patients, then one should not endorse a restriction to stopping that treatment only with terminally ill patients.

Let me turn to a second response to my suggestions that it is a broad "but for" sense of causality that is relevant to moral responsibility for outcomes, and that on this account physicians *do* cause death when they stop life-sustaining food and water or respirators. Consider the legal inquiry into the patient's cause of death. In the case of stopping a life-sustaining respirator, the cause of death would commonly be identified as the underlying disease which resulted in death once the patient was taken off the respirator. Doesn't this suggest that I am mistaken in identifying the physician and what he does as causing death? I think not, and that is because the inquiry into the cause of death is somewhat more complex than it may look at first. It is not simply an empirical inquiry into what conditions played a causal role in the death. It is in part an empirical or factual inquiry of this sort, but it is shaped as well by normative concerns; these are legal concerns when it is a legal inquiry, moral concerns when it is a moral inquiry.

The inquiry into the cause of death is, roughly, something like this: We take all the factors which are "but for" causes, but for these factors the patient would not have died in that way and at that time. There will be a great many such factors, and we do not actually assemble them all, but rather restrict our selection of the cause from among them. On what basis do we select *the* cause of death from among all the "but for" causes? We do not do it solely on empirical or factual grounds; we also consider normative grounds. We ask, for example, in the legal inquiry: Among the "but for" causes, is there anyone who acted in a legally prohibited role with whom the law then wants to concern itself? And we can ask an analogous question in a moral inquiry: Among the "but for" causes, is there anyone whose action was morally impermissible? This helps to explain "cause of death talk" in our earlier two cases in which respirators were stopped by a physician and by a greedy nephew. The action of each is a "but for" cause of the patient's death. The physician's action, however, is within a legally protected role. He acts in accordance with a competent patient's right to refuse even life-sustaining treatment, and so the law does not single him out for further concern when he stops the respirator. The greedy nephew, on the other hand, acts in a legally prohibited role, since (among other things) he acted without the patient's consent, and so the law is concerned further with him. This difference is reflected in our holding the greedy nephew to be the cause of death, while in the case of the physician who also stopped a life-sustaining respirator, the patient's underlying disease is held to be the cause of death.[6]

If this is correct, then it should be no surprise that there is uncertainty and controversy about whether the physician who withholds food and water is the cause of the patient's death. That is a reflection of the uncertainty and controversy that exists about whether stopping food and water is legally and/or morally permissible. If we reach greater agreement on those questions, then it will become clearer whether the physician who stops food and water should be held to be *the* cause of the patient's death. But then it will be the legal and moral permissibility helping determine what or who is *the* cause of death, not whether the physician is the cause being a determinant of legal and moral permissibility.

III. "ORDINARY" AND "EXTRAORDINARY" TREATMENT

I turn finally to the distinction between ordinary and extraordinary or heroic treatment. Those who employ this distinction usually understand the provision of extraordinary treatment as optional and the provision of ordinary treatment as obligatory. This distinction is most often employed in the case of incompetent patients unable to decide for themselves about treatment. And it may seem especially important to the stopping of food and water, since if any care is ordinary, food and water would seem to be. But what is the ordinary/ extraordinary difference? Some understand it to be the degree to which the treatment is (statistically) usual or unusual; others, the degree of invasiveness of treatment; others, the extent to which sophisticated, high technology or artificial treatment is employed; and so forth.

None of these differences, however, are in themselves morally important differences between treatments which could justify distinguishing some as morally obligatory, others as optional. If we understand extraordinary treatments instead as those excessively burdensome or without benefit to the patient, then we do have a difference of obvious moral importance (and, I believe, the meaning the distinction had in its origin within Catholic moral theology). But to determine whether any treatment is excessively burdensome, those burdens must be weighed against the benefits of the treatment in order to judge whether the burdens are worth undergoing. In this interpretation, whether a treatment is ordinary or extraordinary is determined by an assessment of its benefits and burdens. And then, contrary to initial appearances, even providing food and water may be extraordinary care in those few cases in which the burdens of doing so exceed the benefits.

The point I want to press is that the classification of treatments as ordinary or extraordinary now is doing no work in the reasoning about whether the treatment may be withheld. What is doing all the work in the reasoning is the assessment of the benefits and burdens of the treatment. Only when that assessment has been made, and the burdens judged excessive given the benefits, do we label the treatment extraordinary. We put the label on after the analysis has been completed, and so the labeling of treatments as ordinary and extraordinary adds nothing to the analysis—except confusion. It adds confusion because the difference is so often understood not in terms of the benefits and burdens of treatment, but in one of the senses I noted earlier such as the usualness or artificiality of treatment. Given these multiple understandings, and the fact that the ordinary/extraordinary difference adds nothing of substance to the reasoning about forgoing life-sustaining treatment, I believe we do best to avoid its use.

CONCLUSIONS

I will conclude with two points. The first is that I believe the ordinary/extraordinary distinction in its "excessively burdensome" or "without benefit" interpretation does at least point to the correct reasoning in decisions about forgoing food and water. It directs us to assess whether the benefits to a patient from continuing food and water outweigh the burdens, given that patient's particular circumstances. It is my view that there are at least some few cases where it may be that the burdens do outweigh the benefits.

Second, I want to emphasize that I have focused only on a few of the moral complexities and confusions about forgoing life-sustaining treatment in general and about forgoing life-sustaining food and water in particular. There are other difficult moral issues involved, such as the relevance of the economic cost and of the effects on persons other than the patient of feeding or forgoing feeding, which I have not addressed. Moreover, sound public policy should reflect additional considerations, such as slippery slope worries about abuse, the symbolic importance of providing food and water, and how authorization to deny it may affect our care of other potentially vulnerable populations. My concern here has been only with a few of the underlying moral considerations which ought to inform public policy in this area.

NOTES

1 I discuss and defend the assumptions in this argument from example further in "Taking Human Life" [Chapter 5 of this volume], from which my example is drawn. Other arguments to the same effect can be found in James Rachels, "Active and Passive Euthanasia," *N. Engl. J. Med.* 292:75–80 (1975); Michael Tooley, "Defense of Abortion and Infanticide," in *The Problem of Abortion*, Joel Feinberg, ed., Belmont, CA: Wadsworth Publishing Co. (1973), at 84–86; and Jonathan Bennett, "Morality and Consequences," in *The Tanner Lectures on Human Value II*, S. M. McMurrin, ed., Salt Lake City: University of Utah Press (1981), at 45–116.
2 The Appellate Court in *Conroy* tried to rely on such a distinction. *In re Conroy*, 464 A. 2d 303, 315 (N.J. Super. A.D. 1983).
3 Whether tube feedings are part of expected care is an unsettled issue. See results of a survey of physicians in J. Lynn, ed., *By No Extraordinary Means: The Choice to Forgo Food and Water*, Bloomington, IN: Indiana University Press (1986), chapter 4. See also an Australian perspective, W. J. Quilty, "Ethics of

Extraordinary Nutritional Support," *J. Am. Geriatr. Soc.* 32:12–13 (1984).

4 See, e.g., *In re Conroy,* 90 N.J. Super. 453, 464 A.2d 303 (1983), *rev'd,* 98 N.J. 321, 486 A.2d 1209 (1985).

5 See, e.g., Lynn, ed., *By No Extraordinary Means,* chapter 26.

6 *In re Bartling,* 209 Cal. Rptr. 220 (1984). *In re Quinlan,* 70 N.J. 10, 355 A.2d 647, *cert. denied,* 429 U.S. 922 (1976); and *Superintendent of Belchertown State School v. Saikewicz,* 373 Mass. 728, 370 N.E.2d 417 (1977).

Chapter 8

Voluntary active euthanasia

Since the case of Karen Quinlan first seized public attention fifteen years ago, no issue in biomedical ethics has been more prominent than the debate about forgoing life-sustaining treatment. Controversy continues regarding some aspects of that debate, such as forgoing life-sustaining nutrition and hydration, and relevant law varies some from state to state. Nevertheless, I believe it is possible to identify an emerging consensus that competent patients, or the surrogates of incompetent patients, should be permitted to weigh the benefits and burdens of alternative treatments, including the alternative of no treatment, according to the patient's values, and either to refuse any treatment or to select from among available alternative treatments. This consensus is reflected in bioethics scholarship, in reports of prestigious bodies such as the President's Commission for the Study of Ethical Problems in Medicine, The Hastings Center, and the American Medical Association, in a large body of judicial decisions in courts around the country, and finally in the beliefs and practices of health care professionals who care for dying patients.[1]

Earlier versions of this essay were presented at the American Philosophical Association Central Division meetings (at which David Velleman provided extremely helpful comments), Massachusetts General Hospital, Yale University School of Medicine, Princeton University, Brown University, and as the Brin Lecture at The Johns Hopkins School of Medicine. I am grateful to the audiences on each of these occasions, to several anonymous reviewers, and to Norman Daniels for helpful comments. The essay was completed while I was a Fellow in the Program in Ethics and the Professions at Harvard University.

More recently, significant public and professional attention has shifted from life-sustaining treatment to euthanasia – more specifically, voluntary active euthanasia – and to physician-assisted suicide. Several factors have contributed to the increased interest in euthanasia. In the Netherlands, it has been openly practiced by physicians for several years with the acceptance of the country's highest court.[2] In 1988 there was an unsuccessful attempt to get the question of whether it should be made legally permissible on the ballot in California. In November 1991 voters in the state of Washington defeated a widely publicized referendum proposal to legalize both voluntary active euthanasia and physician-assisted suicide. Finally, some cases of this kind, such as "It's Over, Debbie," described in the *Journal of the American Medical Association,* the "suicide machine" of Dr. Jack Kevorkian, and the cancer patient "Diane" of Dr. Timothy Quill, have captured wide public and professional attention.[3] Unfortunately, the first two of these cases were sufficiently problematic that even most supporters of euthanasia or assisted suicide did not defend the physicians' actions in them. As a result, the subsequent debate they spawned has often shed more heat than light. My aim is to increase the light, and perhaps as well to reduce the heat, on this important subject by formulating and evaluating the central ethical arguments for and against voluntary active euthanasia and physician-assisted suicide. My evaluation of the arguments leads me, with reservations to be noted, to support permitting both practices. My primary aim, however, is not to argue for euthanasia, but to identify confusions in some common arguments, and problematic assumptions and claims that need more defense or data in others. The issues are considerably more complex than either supporters or opponents often make out; my hope is to advance the debate by focusing attention on what I believe the real issues under discussion should be.

In the recent bioethics literature some have endorsed physician-assisted suicide but not euthanasia.[4] Are they sufficiently different that the moral arguments for one often do not apply to the other? A paradigm case of physician-assisted suicide is a patient's ending his or her life with a lethal dose of a medication requested of and provided by a physician for that purpose. A paradigm case of voluntary active euthanasia is a physician's administering the lethal dose, often because the patient is unable to do so. The only difference that need exist between the two is the person who actually administers the lethal dose – the physician or the patient. In each, the physician plays an active and necessary causal role.

In physician-assisted suicide the patient acts last (for example, Janet Adkins herself pushed the button after Dr. Kevorkian hooked her up to his suicide machine), whereas in euthanasia the physician acts last by performing the physical equivalent of pushing the button. In both cases, however, the choice rests fully with the patient. In both the patient acts last in the sense of retaining the right to change his or her mind until the point at which the lethal process becomes irreversible. How could there be a substantial moral difference between the two based only on this small difference in the part played by the physician in the causal process resulting in death? Of course, it might be held that the moral difference is clear and important – in euthanasia the physician kills the patient whereas in physician-assisted suicide the patient kills him- or herself. But this is misleading at best. In assisted suicide the physician and patient together kill the patient. To see this, suppose a physician supplied a lethal dose to a patient with the knowledge and intent that the patient will wrongfully administer it to another. We would have no difficulty in morality or the law recognizing this as a case of joint action to kill for which both are responsible.

If there is no significant, intrinsic moral difference between the two, it is also difficult to see why public or legal policy should permit one but not the other; worries about abuse or about giving anyone dominion over the lives of others apply equally to either. As a result, I will take the arguments evaluated below to apply to both and will focus on euthanasia.

My concern here will be with *voluntary* euthanasia only – that is, with the case in which a clearly competent patient makes a fully voluntary and persistent request for aid in dying. Involuntary euthanasia, in which a competent patient explicitly refuses or opposes receiving euthanasia, and non-voluntary euthanasia, in which a patient is incompetent and unable to express his or her wishes about euthanasia, will be considered here only as potential unwanted side effects of permitting voluntary euthanasia. I emphasize as well that I am concerned with *active* euthanasia, not withholding or withdrawing life-sustaining treatment, which some commentators characterize as "passive euthanasia." Finally, I will be concerned with euthanasia where the motive of those who perform it is to respect the wishes of the patient and to provide the patient with a "good death," though one important issue is whether a change in legal policy could restrict the performance of euthanasia to only those cases.

A last introductory point is that I will be examining only secular arguments about euthanasia, though of course many people's attitudes to it are inextricable from their religious views. The policy issue is only whether euthanasia should be permissible, and no one who has religious objections to it should be required to take any part in it, though of course this would not fully satisfy some opponents.

THE CENTRAL ETHICAL ARGUMENT FOR VOLUNTARY ACTIVE EUTHANASIA

The central ethical argument for euthanasia is familiar. It is that the very same two fundamental ethical values supporting the consensus on patients' rights to decide about life-sustaining treatment also support the ethical permissibility of euthanasia. These values are individual self-determination or autonomy and individual well-being. By self-determination as it bears on euthanasia, I mean people's interest in making important decisions about their lives for themselves according to their own values or conceptions of a good life, and in being left free to act on those decisions. Self-determination is valuable because it permits people to form and live in accordance with their own conception of a good life, at least within the bounds of justice and consistent with others doing so as well. In exercising self-determination people take responsibility for their lives and for the kinds of persons they become. A central aspect of human dignity lies in people's capacity to direct their lives in this way. The value of exercising self-determination presupposes some minimum of decisionmaking capacities or competence, which thus limits the scope of euthanasia supported by self-determination; it cannot justifiably be administered, for example, in cases of serious dementia or treatable clinical depression.

Does the value of individual self-determination extend to the time and manner of one's death? Most people are very concerned about the nature of the last stage of their lives. This reflects not just a fear of experiencing substantial suffering when dying, but also a desire to retain dignity and control during this last period of life. Death is today increasingly preceded by a long period of significant physical and mental decline, due in part to the technological interventions of modern medicine. Many people adjust to these disabilities and find meaning and value in new activities and ways. Others find the impairments and burdens in the last stage of their lives at some point sufficiently great to make life no longer worth living. For many patients near death, maintaining the quality of one's life, avoiding great suffering, maintaining one's dignity, and insuring that others remember us as we wish them to become of paramount importance and outweigh merely extending one's life. But there is no single, objectively correct answer for everyone as to when, if at all, one's life becomes all things considered a burden and unwanted. If self-determination is a fundamental value, then the great variability among people on this question makes it especially important that individuals control the manner, circumstances, and timing of their dying and death.

The other main value that supports euthanasia is individual well-being. It might seem that individual well-being conflicts with a person's self-determination when the person requests euthanasia. Life itself is commonly taken to be a central good for persons, often valued for its own sake, as well as necessary for pursuit of all other goods within a life. But when a competent patient decides to forgo all further life-sustaining treatment then the patient, either explicitly or implicitly, commonly decides that the best life possible for him or her with treatment is of sufficiently poor quality that it is worse than no further life at all. Life is no longer considered a benefit by the patient, but has now become a burden. The same judgment underlies a request for euthanasia: continued life is seen by the patient as no longer a benefit, but now a burden. Especially in the often severely compromised and debilitated states of many critically ill or dying patients, there is no objective standard, but only the competent patient's judgment of whether continued life is no longer a benefit.

Of course, sometimes there are conditions, such as clinical depression, that call into question whether the patient has made a competent choice, either to forgo life-sustaining treatment or to seek euthanasia, and then the patient's choice need not be evidence that continued life is no longer a benefit for him or her. Just as with decisions about treatment, a determination of incompetence can warrant not honoring the patient's choice; in the case of treatment, we then transfer decisional authority to a surrogate, though in the case of voluntary active euthanasia a determination that the patient is incompetent means that choice is not possible.

The value or right of self-determination does not entitle patients to compel physicians to act contrary to their own moral or professional values. Physicians are moral and professional agents whose own self-determination or integrity should be respected as well. If performing euthanasia became legally permissible, but conflicted with a particular physician's reasonable understanding of his or her moral or professional responsibilities, the care of a patient who requested euthanasia should be transferred to another.

Most opponents do not deny that there are some cases in which the values of patient self-determination and well-being support euthanasia. Instead, they commonly offer two kinds of arguments against it that on their view outweigh or override this support. The first kind of argument is that in any individual case where considerations of the patient's self-determination and well-being do support euthanasia, it is nevertheless always ethically wrong or impermissible. The second kind of argument grants that in some individual cases euthanasia may *not* be ethically wrong, but maintains nonetheless that public and legal policy should never permit it. The first kind of argument focuses on features of any individual case of euthanasia, while the second kind focuses on social or legal policy. In the next section I consider the first kind of argument.

EUTHANASIA IS THE DELIBERATE KILLING OF AN INNOCENT PERSON

The claim that any individual instance of euthanasia is a case of deliberate killing of an innocent person is, with only minor qualifications, correct. Unlike forgoing life-sustaining treatment, commonly understood as allowing to die, euthanasia is clearly killing, defined as depriving of life or causing the death of a living being. While providing morphine for pain relief at doses where the risk of respiratory depression and an earlier death may be a foreseen but unintended side effect of treating the patient's pain, in a case of euthanasia the patient's death is deliberate or intended even if in both the physician's ultimate end may be respecting the patient's wishes. If the deliberate killing of an innocent person is wrong, euthanasia would be nearly always impermissible.

In the context of medicine, the ethical prohibition against deliberately killing the innocent derives some of its plausibility from the belief that nothing in the currently accepted practice of medicine is deliberate killing. Thus, in commenting on the "It's Over, Debbie" case, four prominent physicians and bioethicists could entitle their paper "Doctors Must Not Kill."[5] The belief that doctors do not in fact kill requires the corollary belief that forgoing life-sustaining treatment, whether by not starting or by stopping treatment, is allowing to die, not killing. Common though this view is, I shall argue that it is confused and mistaken.

Why is the common view mistaken? Consider the case of a patient terminally ill with ALS disease. She is completely respirator dependent with no hope of ever being weaned. She is unquestionably competent but finds her condition intolerable and persistently requests to be removed from the respirator and allowed to die. Most people and physicians would agree that the patient's physician should respect the patient's wishes and remove her from the respirator, though this will certainly cause the patient's death. The common understanding is that the physician thereby allows the patient to die. But is that correct?

Suppose the patient has a greedy and hostile son who mistakenly believes that his mother will never decide to stop her life-sustaining treatment and that even if she did her physician would not remove her from the respirator. Afraid that his inheritance will be dissipated by a long and expensive hospitalization, he enters his mother's room while she is sedated, extubates her, and she dies. Shortly thereafter the medical staff discovers what he has done and confronts the son. He replies, "I didn't kill her, I merely allowed her to die. It was her ALS disease that caused her death." I think this would rightly be dismissed as transparent sophistry – the son went into his mother's room and deliberately killed her. But, of course, the son performed just the same physical actions, did just the same thing, that the physician would have done. If that is so, then doesn't the physician also kill the patient when he extubates her?

I underline immediately that there are important ethical differences between what the physician and the greedy son do. First, the physician acts with the patient's consent whereas the son does not. Second, the physician acts with a good motive – to respect the patient's wishes and self-determination – whereas the son acts with a bad motive – to protect his own inheritance. Third, the physician acts in a social role through which he is legally authorized to carry out the patient's wishes regarding treatment whereas the son

209

has no such authorization. These and perhaps other ethically important differences show that what the physician did was morally justified whereas what the son did was morally wrong. What they do *not* show, however, is that the son killed while the physician allowed to die. One can either kill or allow to die with or without consent, with a good or bad motive, within or outside of a social role that authorizes one to do so.

The difference between killing and allowing to die that I have been implicitly appealing to here is roughly that between acts and omissions resulting in death.[6] Both the physician and the greedy son act in a manner intended to cause death, do cause death, and so both kill. One reason this conclusion is resisted is that on a different understanding of the distinction between killing and allowing to die, what the physician does is allow to die. In this account, the mother's ALS is a lethal disease whose normal progression is being held back or blocked by the life-sustaining respirator treatment. Removing this artificial intervention is then viewed as standing aside and allowing the patient to die of her underlying disease. I have argued elsewhere that this alternative account is deeply problematic, in part because it commits us to accepting that what the greedy son does is to allow to die, not kill.[7] Here, I want to note two other reasons why the conclusion that stopping life support is killing is resisted.

The first reason is that killing is often understood, especially within medicine, as unjustified causing of death; in medicine it is thought to be done only accidentally or negligently. It is also increasingly widely accepted that a physician is ethically justified in stopping life support in a case like that of the ALS patient. But if these two beliefs are correct, then what the physician does cannot be killing, and so must be allowing to die. Killing patients is not, to put it flippantly, understood to be part of physicians' job description. What is mistaken in this line of reasoning is the assumption that all killings are *unjustified* causings of death. Instead, some killings are ethically justified, including many instances of stopping life support.

Another reason for resisting the conclusion that stopping life support is often killing is that it is psychologically uncomfortable. Suppose the physician had stopped the ALS patient's respirator and had made the son's claim, "I didn't kill her, I merely allowed her to die. It was her ALS disease that caused her death." The clue to the psychological role here is how naturally the "merely" modifies "allowed her to die." The characterization as allowing to die is meant to shift felt responsibility away from the agent – the physician – and to the lethal disease process. Other language common in death and dying contexts plays a similar role; "letting nature take its course" or "stopping prolonging the dying process" both seem to shift responsibility from the physician who stops life support to the fatal disease process. However psychologically helpful these conceptualizations may be in making the difficult responsibility of a physician's role in the patient's death bearable, they nevertheless are confusions. Both physicians and family members can instead be helped to understand that it is the patient's decision and consent to stopping treatment that limits their responsibility for the patient's death and that shifts that responsibility to the patient.

Many who accept the difference between killing and allowing to die as the distinction between acts and omissions resulting in death have gone on to argue that killing is not in itself morally different from allowing to die.[8] In this account, very roughly, one kills when one performs an action that causes the death of a person (we are in a boat, you cannot swim, I push you overboard, and you drown), and one allows to die when one has the ability and opportunity to prevent the death of another, knows this, and omits doing so, with the result that the person dies (we are in a boat, you cannot swim, you fall overboard, I don't throw you an available life ring, and you drown). Those who see no moral difference between killing and allowing to die typically employ the strategy of comparing cases that differ in these and no other potentially morally important respects. This will allow people to consider whether the mere difference that one is a case of killing and the other of allowing to die matters morally, or whether instead it is other features that make most cases of killing worse than most instances of allowing to die. Here is such a pair of cases:

Case 1. A very gravely ill patient is brought to a hospital emergency room and sent up to the ICU. The patient begins to develop respiratory failure that is likely to require intubation very soon. At that point the patient's family members and long-standing physician arrive at the ICU and inform the ICU staff that there had been extensive discussion about future care with the patient when he was unquestionably competent. Given his grave and terminal illness, as well as his state of debilitation, the patient had firmly rejected being placed on a respirator under any circumstances, and the family and physician produce the patient's advance directive to that effect. The ICU staff do not intubate the patient, who dies of respiratory failure.

Case 2. The same as Case 1 except that the family and physician are slightly delayed in traffic and arrive shortly after the patient has been intubated and placed on the respirator. The ICU staff extubate the patient, who dies of respiratory failure.

In Case 1 the patient is allowed to die, in Case 2 he is killed, but it is hard to see why what is done in Case 2 is significantly different morally than what is done in Case 1. It must be other factors that make most killings worse than most allowings to die, and if so, euthanasia cannot be wrong simply because it is killing instead of allowing to die.

Suppose both my arguments are mistaken. Suppose that killing is worse than allowing to die and that withdrawing life support is not killing, although euthanasia is. Euthanasia still need not for that reason be morally wrong. To see this, we need to determine the basic principle for the moral evaluation of killing persons. What is it that makes paradigm cases of wrongful killing wrongful? One very plausible answer is that killing denies the victim something that he or she values greatly – continued life or a future. Moreover, since continued life is necessary for pursuing any of a person's plans and purposes, killing brings the frustration of all of these plans and desires as well. In a nutshell, wrongful killing deprives a person of a valued future, and of all the person wanted and planned to do in that future.

A natural expression of this account of the wrongness of killing is that people have a moral right not to be killed.[9] But in this account of the wrongness of killing, the right not to be killed, like other rights, should be waivable when the person makes a competent decision that continued life is no longer wanted or a good, but is instead worse than no further life at all. In this view, euthanasia is properly understood as a case of a person having waived his or her right not to be killed.

This rights view of the wrongness of killing is not, of course, universally shared. Many people's moral views about killing have their origins in religious views that human life comes from God and cannot be justifiably destroyed or taken away, either by the person whose life it is or by another. But in a pluralistic society like our own with a strong commitment to freedom of religion, public policy should not be grounded in religious beliefs which many in that society reject. I turn now to the general evaluation of public policy on euthanasia.

WOULD THE BAD CONSEQUENCES OF EUTHANASIA OUTWEIGH THE GOOD?

The argument against euthanasia at the policy level is stronger than at the level of individual cases, though even here I believe the case is ultimately unpersuasive, or at best indecisive. The policy level is the place where the main issues lie, however, and where moral considerations that might override arguments in favor of euthanasia will be found, if they are found anywhere. It is important to note two kinds of disagreement about the consequences for public policy of permitting euthanasia. First, there is empirical or factual disagreement about what the consequences would be. This disagreement is greatly exacerbated by the lack of firm data on the issue. Second, since on any reasonable assessment there would be both good and bad consequences, there are moral disagreements about the relative importance of different effects. In addition to these two sources of disagreement, there is also no single, well-specified policy proposal for legalizing euthanasia on which policy assessments can focus. But without such specification, and especially without explicit procedures for protecting against well-intentioned misuse and ill-intentioned abuse, the consequences for policy are largely speculative. Despite these difficulties, a preliminary account of the main likely good and bad consequences is possible. This should help clarify where better data or more moral analysis and argument are needed, as well as where policy safeguards must be developed.

Potential good consequences of permitting euthanasia

What are the likely good consequences? First, if euthanasia were permitted it would be possible to respect the self-determination of competent patients who want it, but now cannot get it because of its illegality. We simply do not know how many such patients and people there are. In the Netherlands, with a population of about 14.5 million (in 1987), estimates in a recent study were that about 1,900 cases of voluntary active euthanasia or physician-assisted suicide occur annually. No straightforward extrapolation to the United States is possible for many reasons, among them, that we do not know how many people here who want euthanasia now get it, despite its illegality. Even with better data on the number of persons who want euthanasia but cannot get it, significant moral disagreement would remain about how much weight should be given to any instance of failure to respect a person's self-determination in this way.

One important factor substantially affecting the number of persons who would seek euthanasia is the extent to which an alternative is available. The widespread acceptance in the law, social policy, and medical practice of the right of a competent patient to forgo life-sustaining treatment suggests that the number of competent persons in the United States who would want euthanasia if it were permitted is probably relatively small.

A second good consequence of making euthanasia legally permissible benefits a much larger group. Polls have shown that a majority of the American public believes that people should have a right to obtain euthanasia if they want it.[10] No doubt the vast majority of those who support this right to euthanasia will never in fact come to want euthanasia for themselves. Nevertheless, making it legally permissible would reassure many people that if they ever do want euthanasia they would be able to obtain it. This reassurance would supplement the broader control over the process of dying given by the right to decide about life-sustaining treatment. Having fire insurance on one's house benefits all who have it, not just those whose houses actually burn down, by reassuring them that in the unlikely event of their house burning down, they will receive the money needed to rebuild it. Likewise, the legalization of euthanasia can be thought of as a kind of insurance policy against being forced to endure a protracted dying process that one has come to find burdensome and unwanted, especially when there is no life-sustaining treatment to forgo. The strong concern about losing control of their care expressed by many people who face serious illness likely to end in death suggests that they give substantial importance to the legalization of euthanasia as a means of maintaining this control.

A third good consequence of the legalization of euthanasia concerns patients whose dying is filled with severe and unrelievable pain of suffering. When there is a life-sustaining treatment that, if forgone, will lead relatively quickly to death, then doing so can bring an end to these patients' suffering without recourse to euthanasia. For patients receiving no such treatment, however, euthanasia may be the only release from their otherwise prolonged suffering and agony. This argument from mercy has always been the strongest argument for euthanasia in those cases to which it applies.[11]

The importance of relieving pain and suffering is less controversial than is the frequency with which patients are forced to undergo untreatable agony that only euthanasia could relieve. If we focus first on suffering caused by physical pain, it is crucial to distinguish pain that *could* be adequately relieved with modern methods of pain control, though it in fact is not, from pain that is relievable only by death.[12] For a variety of reasons, including some physicians' fear of hastening the patient's death, as well as the lack of a publicly accessible means for assessing the amount of the patient's pain, many patients suffer pain that could be, but is not, relieved.

Specialists in pain control, as for example the pain of terminally ill cancer patients, argue that there are very few patients whose pain could not be adequately controlled, though sometimes at the cost of so sedating them that they are effectively unable to interact with other people or their environment. Thus, the argument from mercy in cases of physical pain can probably be met in a large majority of cases by providing adequate measures of pain relief. This should be a high priority, whatever our legal policy on euthanasia – the relief of pain and suffering has long been, quite properly, one of the central goals of medicine. Those cases in which pain could be effectively relieved, but in fact is not, should only count significantly in favor of legalizing euthanasia if all reasonable efforts to change pain management techniques have been tried and have failed.

Dying patients often undergo substantial psychological suffering that is not fully or even principally the result of physical pain.[13] The knowledge about how to relieve this suffering is much more limited than in the case of relieving pain, and efforts to do so are probably more often unsuccessful. If the argument from mercy is extended to patients experiencing great and unrelievable psychological suffering, the numbers of patients to which it applies are much greater.

One last good consequence of legalizing euthanasia is that

once death has been accepted, it is often more humane to end life quickly and peacefully, when that is what the patient wants. Such a death will often be seen as better than a more prolonged one. People who suffer a sudden and unexpected death, for example by dying quickly or in their sleep from a heart attack or stroke, are often considered lucky to have died in this way. We care about how we die in part because we care about how others remember us, and we hope they will remember us as we were in "good times" with them and not as we might be when disease has robbed us of our dignity as human beings. As with much in the treatment and care of the dying, people's concerns differ in this respect, but for at least some people, euthanasia will be a more humane death than what they have often experienced with other loved ones and might otherwise expect for themselves.

Some opponents of euthanasia challenge how much importance should be given to any of these good consequences of permitting it, or even whether some would be good consequences at all. But more frequently, opponents cite a number of bad consequences that permitting euthanasia would or could produce, and it is to their assessment that I now turn.

Potential bad consequences of permitting euthanasia

Some of the arguments against permitting euthanasia are aimed specifically against physicians, while others are aimed against anyone being permitted to perform it. I shall first consider one argument of the former sort. Permitting physicians to perform euthanasia, it is said, would be incompatible with their fundamental moral and professional commitment as healers to care for patients and to protect life. Moreover, if euthanasia by physicians became common, patients would come to fear that a medication was intended not to treat or care, but instead to kill, and would thus lose trust in their physicians. This position was forcefully stated in a paper by Willard Gaylin and his colleagues:

The very soul of medicine is on trial. . . . This issue touches medicine at its moral center; if this moral center collapses, if physicians become killers or are even licensed to kill, the profession – and, therewith, each physician – will never again be worthy of trust and respect as healer and comforter and protector of life in all its frailty.

These authors go on to make clear that, while they oppose permitting anyone to perform euthanasia, their special concern is with physicians doing so:

We call on fellow physicians to say that they will not deliberately kill. We must also say to each of our fellow physicians that we will not tolerate killing of patients and that we shall take disciplinary action against doctors who kill. And we must say to the broader community that if it insists on tolerating or legalizing active euthanasia, it will have to find nonphysicians to do its killing.[14]

If permitting physicians to kill would undermine the very "moral center" of medicine, then almost certainly physicians should not be permitted to perform euthanasia. But how persuasive is this claim? Patients should not fear, as a consequence of permitting *voluntary* active euthanasia, that their physicians will substitute a lethal injection for what patients want and believe is part of their care. If active euthanasia is restricted to cases in which it is truly voluntary, then no patient should fear getting it unless she or he has voluntarily requested it. (The fear that we might in time also come to accept nonvoluntary, or even involuntary, active euthanasia is a slippery slope worry I address below.) Patients' trust of their physicians could be increased, not eroded, by knowledge that physicians will provide aid in dying when patients seek it.

Might Gaylin and his colleagues nevertheless be correct in their claim that the moral center of medicine would collapse if physicians were to become killers? This question raises what at the deepest level should be the guiding aims of medicine, a question that obviously cannot be fully explored here. But I do want to say enough to indicate the direction that I believe an appropriate response to this challenge should take. In spelling out above what I called the positive argument for voluntary active euthanasia, I suggested that two principal values – respecting patients' self-determination and promoting their well-being – underlie the consensus that competent patients, or the surrogates of incompetent patients, are entitled to refuse any life-sustaining treatment and to choose from among available alternative treatments. It is the commitment to these two values in guiding physicians' actions as healers, comforters, and protectors of their patients' lives that should be at the "moral center" of medicine, and these two values support physicians' administering euthanasia when their patients make competent requests for it.

What should not be at that moral center is a commitment to preserving patients' lives as such, without regard to whether those patients want their lives preserved or judge their preservation a benefit to them. Vitalism has been rejected by most physicians, and despite some statements that suggest it, is almost certainly not what Gaylin and colleagues intended. One of them, Leon Kass, has elaborated elsewhere the view that medicine is a moral profession whose proper aim is "the naturally given end of health," understood as the wholeness and well-working of the human being; "for the physician, at least, human life in living bodies commands respect and reverence – *by its very nature*." Kass continues, "the deepest ethical principle restraining the physician's power is not the autonomy or freedom of the patient; neither is it his own compassion or good intention. Rather, it is the dignity and mysterious power of human life itself."[15] I believe Kass is in the end mistaken about the proper account of the aims of medicine and the limits on physicians' power, but this difficult issue will certainly be one of the central themes in the continuing debate about euthanasia.

A second bad consequence that some foresee is that permitting euthanasia would weaken society's commitment to provide optimal care for dying patients. We live at a time in which the control of health care costs has become, and is likely to continue to be, the dominant focus of health care policy. If euthanasia is seen as a cheaper alternative to adequate care and treatment, then we might become less scrupulous about providing sometimes costly support and other services to dying patients. Particularly if our society comes to embrace deeper and more explicit rationing of health care, frail, elderly, and dying patients will need to be strong and effective advocates for their own health care and other needs, although they are hardly in a position to do this. We should do nothing to weaken their ability to obtain adequate care and services.

This second worry is difficult to assess because there is little firm evidence about the likelihood of the feared erosion in the care of dying patients. There are at least two reasons, however, for skepticism about this argument. The first is that the same worry could have been directed at recognizing patients' or surrogates' rights to forgo life-sustaining treatment, yet there is no persuasive evidence that recognizing the right to refuse treatment has caused a serious erosion in the quality of care of dying patients. The second reason for skepticism about this worry is that only a very small proportion of deaths would occur from euthanasia if it were permitted. In the Netherlands, where euthanasia under specified circumstances is permitted by the courts, though not authorized by statute, the best estimate of the proportion of overall deaths that result from it is about 2 percent.[16] Thus, the vast majority of critically ill and dying patients will not request it, and so will still have to be cared for by physicians, families, and others. Permitting euthanasia should not diminish people's commitment and concern to maintain and improve the care of these patients.

A third possible bad consequence of permitting euthanasia (or even a public discourse in which strong support for euthanasia is evident) is to threaten the progress made in securing the rights of patients or their surrogates to decide about and to refuse life-sustaining treatment.[17] This progress has been made against the backdrop of a clear and firm legal prohibition of euthanasia, which has provided a relatively bright line limiting the dominion of others over patients' lives. It has therefore been an important reassurance to concerns about how the authority to take steps ending life might be misused, abused, or wrongly extended.

Many supporters of the right of patients or their surrogates to refuse treatment strongly oppose euthanasia, and if forced to choose might well withdraw their support of the right to refuse treatment rather than accept euthanasia. Public policy in the last fifteen years has generally let life-sustaining treatment decisions be made in health care settings between physicians and patients or their surrogates, and without the involvement of the courts. However, if euthanasia is made legally permissible greater involvement of the courts is likely, which could in turn extend to a greater court involvement in life-sustaining treatment decisions. Most agree, however, that increased involvement of the courts in these decisions would be undesirable, as it would make sound decision-making more cumbersome and difficult without sufficient compensating benefits.

As with the second potential bad consequence of permitting euthanasia, this third consideration too is speculative and difficult to assess. The feared erosion of patients' or surrogates' rights to decide about life-sustaining treatment, together with greater court involvement in those decisions, are both possible. However, I believe there is reason to discount this general worry. The legal rights of competent patients and, to a lesser degree, surrogates of incompetent patients to decide about treatment are very firmly embedded in a long line of informed consent and life-sustaining treatment cases, and are not likely to be eroded by a debate over, or even acceptance of, euthanasia. It will not be accepted without safeguards that reassure the public about abuse, and if that debate shows the need for similar safeguards for some life-sustaining treatment decisions they should be adopted there as well. In neither case are the only possible safeguards greater court involvement, as the recent growth of institutional ethics committees shows.

The fourth potential bad consequence of permitting euthanasia has been developed by David Velleman and turns on the subtle point that making a new option or choice available to people can sometimes make them worse off, even if once they have the choice they go on to choose what is best for them.[18] Ordinarily, people's continued existence is viewed by them as given, a fixed condition with which they must cope. Making euthanasia available to people as an option denies them the alternative of staying alive by default. If people are offered the option of euthanasia, their continued existence is now a choice for which they can be held responsible and which they can be asked by others to justify. We care, and are right to care, about being able to justify ourselves to others. To the extent that our society is unsympathetic to justifying a severely dependent or impaired existence, a heavy psychological burden of proof may be placed on patients who think their terminal illness or chronic infirmity is not a sufficient reason for dying. Even if they otherwise view their life as worth living, the opinion of others around them that it is not can threaten their reason for living and make euthanasia a rational choice. Thus the existence of the option becomes a subtle pressure to request it.

This argument correctly identifies the reason why offering some patients the option of euthanasia would not benefit them. Velleman takes it not as a reason for opposing all euthanasia, but for restricting it to circumstances where there are "unmistakable and overpowering reasons for persons to want the option of euthanasia," and for denying the option in all other cases. But there are at least three reasons why such restriction may not be warranted. First, polls and other evidence support that most Americans believe euthanasia should be permitted (though the recent defeat of the referendum to permit it in the state of Washington raises some doubt about this support). Thus, many more people seem to want the choice than would be made worse off by getting it. Second, if giving people the option of ending their life really makes them worse off, then we should not only prohibit euthanasia, but also take back from people the right they now have to decide about life-sustaining treatment. The feared harmful effect should already have occurred from securing people's right to refuse life-sustaining treatment, yet there is no evidence of any such widespread harm or any broad public desire to rescind that right. Third, since there is a wide range of conditions in which reasonable people can and do disagree about whether they would want continued life, it is not possible to restrict the permissibility of euthanasia as narrowly as Velleman suggests without thereby denying it to most persons who would want it; to permit it only in cases in which virtually everyone would want it would be to deny it to most who would want it.

A fifth potential bad consequence of making euthanasia legally permissible is that it might weaken the general legal prohibition of homicide. This prohibition is so fundamental to civilized society, it is argued, that we should do nothing that erodes it. If most cases of stopping life support are killing, as I have already argued, then the court cases permitting such killing have already in effect weakened this prohibition. However, neither the courts nor most people have seen these cases as killing and so as challenging the prohibition of homicide. The courts have usually grounded patients' or their surrogates' rights to refuse life-sustaining treatment in rights to privacy, liberty, self-determination, or bodily integrity, not in exceptions to homicide laws.

Legal permission for physicians or others to perform euthanasia could not be grounded in patients' rights to decide about medical treatment. Permitting euthanasia would require qualifying, at least in effect, the legal prohibition against homicide, a prohibition that in general does not allow the consent of the victim to justify or excuse the act. Nevertheless, the very same fundamental basis of the right to decide about life-sustaining treatment – respecting a person's self-determination – does support euthanasia as well. Individual self-determination has long been a well-entrenched and fundamental value in the law, and so extending it to euthanasia would not require appeal to novel legal values or principles. That suicide or attempted suicide is no longer a

criminal offense in virtually all states indicates an acceptance
of individual self-determination in the taking of one's own
life analogous to that required for voluntary active euthanasia. The legal prohibition (in most states) of assisting in
suicide and the refusal in the law to accept the consent of
the victim as a possible justification of homicide are both
arguably a result of difficulties in the legal process of establishing the consent of the victim after the fact. If procedures
can be designed that clearly establish the voluntariness of
the person's request for euthanasia, it would under those
procedures represent a carefully circumscribed qualification
on the legal prohibition of homicide. Nevertheless, some
remaining worries about this weakening can be captured in
the final potential bad consequence, to which I will now turn.

This final potential bad consequence is the central concern
of many opponents of euthanasia and, I believe, is the most
serious objection to a legal policy permitting it. According to
this "slippery slope" worry, although active euthanasia may
be morally permissible in cases in which it is unequivocally
voluntary and the patient finds his or her condition unbearable, a legal policy permitting euthanasia would inevitably
lead to active euthanasia being performed in many other
cases in which it would be morally wrong. To prevent those
other wrongful cases of euthanasia we should not permit
even morally justified performance of it.

Slippery slope arguments of this form are problematic and
difficult to evaluate.[19] From one perspective, they are the last
refuge of conservative defenders of the status quo. When all
the opponent's objections to the wrongness of euthanasia
itself have been met, the opponent then shifts ground and
acknowledges both that it is not in itself wrong and that a
legal policy which resulted only in its being performed would
not be bad. Nevertheless, the opponent maintains, it should
still not be permitted because doing so would result in its
being performed in other cases in which it is not voluntary
and would be wrong. In this argument's most extreme form,
permitting euthanasia is the first and fateful step down the

slippery slope to Nazism. Once on the slope we will be unable to get off.

Now it cannot be denied that it is *possible* that permitting euthanasia could have these fateful consequences, but that cannot be enough to warrant prohibiting it if it is otherwise justified. A similar *possible* slippery slope worry could have been raised to securing competent patients' rights to decide about life support, but recent history shows such a worry would have been unfounded. It must be relevant how likely it is that we will end with horrendous consequences and an unjustified practice of euthanasia. How *likely* and *widespread* would the abuses and unwarranted extensions of permitting it be? By abuses, I mean the performance of euthanasia that fails to satisfy the conditions required for voluntary active euthanasia, for example, if the patient has been subtly pressured to accept it. By unwarranted extensions of policy, I mean later changes in legal policy to permit not just voluntary euthanasia, but also euthanasia in cases in which, for example, it need not be fully voluntary. Opponents of voluntary euthanasia on slippery slope grounds have not provided the data or evidence necessary to turn their speculative concerns into well-grounded likelihoods.

It is at least clear, however, that both the character and likelihood of abuses of a legal policy permitting euthanasia depend in significant part on the procedures put in place to protect against them. I will not try to detail fully what such procedures might be, but will just give some examples of what they might include:

1. The patient should be provided with all relevant information about his or her medical condition, current prognosis, available alternative treatments, and the prognosis of each.
2. Procedures should ensure that the patient's request for euthanasia is stable or enduring (a brief waiting period could be required) and fully voluntary (an advocate for the patient might be appointed to ensure this).

3. All reasonable alternatives must have been explored for improving the patient's quality of life and relieving any pain or suffering.
4. A psychiatric evaluation should ensure that the patient's request is not the result of a treatable psychological impairment such as depression.[20]

These examples of procedural safeguards are all designed to ensure that the patient's choice is fully informed, voluntary, and competent, and so a true exercise of self-determination. Other proposals for euthanasia would restrict its permissibility further – for example, to the terminally ill – a restriction that cannot be supported by self-determination. Such additional restrictions might, however, be justified by concern for limiting potential harms from abuse. At the same time, it is important not to impose procedural or substantive safeguards so restrictive as to make euthanasia impermissible or practically infeasible in a wide range of justified cases.

These examples of procedural safeguards make clear that it is possible to substantially reduce, though not to eliminate, the potential for abuse of a policy permitting voluntary active euthanasia. Any legalization of the practice should be accompanied by a well-considered set of procedural safeguards together with an ongoing evaluation of its use. Introducing euthanasia into only a few states could be a form of carefully limited and controlled social experiment that would give us evidence about the benefits and harms of the practice. Even then firm and uncontroversial data may remain elusive, as the continuing controversy over what has taken place in the Netherlands in recent years indicates.[21]

The slip into nonvoluntary active euthanasia

While I believe slippery slope worries can largely be limited by making necessary distinctions both in principle and in practice, one slippery slope concern is legitimate. There is reason to expect that legalization of voluntary active euthanasia might soon be followed by strong pressure to legalize some nonvoluntary euthanasia of incompetent patients unable to express their own wishes. Respecting a person's self-determination and recognizing that continued life is not always of value to a person can support not only voluntary active euthanasia, but some nonvoluntary euthanasia as well. These are the same values that ground competent patients' right to refuse life-sustaining treatment. Recent history here is instructive. In the medical ethics literature, in the courts since Quinlan, and in norms of medical practice, that right has been extended to incompetent patients and exercised by a surrogate who is to decide as the patient would have decided in the circumstances if competent.[22] It has been held unreasonable to continue life-sustaining treatment that the patient would not have wanted just because the patient now lacks the capacity to tell us that. Life-sustaining treatment for incompetent patients is today frequently forgone on the basis of a surrogate's decision, or less frequently on the basis of an advance directive executed by the patient while still competent. The very same logic that has extended the right to refuse life-sustaining treatment from a competent patient to the surrogate of an incompetent patient (acting with or without a formal advance directive from the patient) may well extend the scope of active euthanasia. The argument will be, Why continue to force unwanted life on patients just because they have now lost the capacity to request euthanasia from us?

A related phenomenon may reinforce this slippery slope concern. In the Netherlands, what the courts have sanctioned has been clearly restricted to voluntary euthanasia. In itself, this serves as some evidence that permitting it need *not* lead to permitting the nonvoluntary variety. There is some indication, however, that for many Dutch physicians euthanasia is no longer viewed as a special action, set apart from their usual practice and restricted only to competent persons.[23] Instead, it is seen as one end of a spectrum of caring for dying patients. When viewed in this way it will be difficult to deny euthanasia to a patient for whom it is seen as the best or most appropriate form of care simply because that patient is now incompetent and cannot request it.

Even if voluntary active euthanasia should slip into non-voluntary active euthanasia, with surrogates acting for incompetent patients, the ethical evaluation is more complex than many opponents of euthanasia allow. Just as in the case of surrogates' decisions to forgo life-sustaining treatment for incompetent patients, so also surrogates' decisions to request euthanasia for incompetent persons would often accurately reflect what the incompetent person would have wanted and would deny the person nothing that he or she would have considered worth having. Making nonvoluntary active euthanasia legally permissible, however, would greatly enlarge the number of patients on whom it might be performed and substantially enlarge the potential for misuse and abuse. As noted above, frail and debilitated elderly people, often demented or otherwise incompetent and thereby unable to defend and assert their own interests, may be especially vulnerable to unwanted euthanasia.

For some people, this risk is more than sufficient reason to oppose the legalization of voluntary euthanasia. But while we should in general be cautious about inferring much from the experience in the Netherlands to what our own experience in the United States might be, there may be one important lesson that we can learn from them. One commentator has noted that in the Netherlands families of incompetent patients have less authority than do families in the United States to act as surrogates for incompetent patients in making decisions to forgo life-sustaining treatment.[24] From the Dutch perspective, it may be we in the United States who are *already* on the slippery slope in having given surrogates broad authority to forgo life-sustaining treatment for incompetent persons. In this view, the more important moral divide, and the more important with regard to potential for abuse, is not between forgoing life-sustaining treatment and euthanasia, but instead between voluntary and nonvoluntary performance of either. If this is correct, then the more important issue is ensuring the appropriate principles and procedural safeguards for the exercise of decisionmaking authority by surrogates for incompetent persons in *all* decisions at the end of life. This may be the correct response to slippery slope worries about euthanasia.

I have cited both good and bad consequences that have been thought likely from a policy change permitting voluntary active euthanasia, and have tried to evaluate their likelihood and relative importance. Nevertheless, as I noted earlier, reasonable disagreement remains both about the consequences of permitting euthanasia and about which of these consequences are more important. The depth and strength of public and professional debate about whether, all things considered, permitting euthanasia would be desirable or undesirable reflects these disagreements. While my own view is that the balance of considerations supports permitting the practice, my principal purpose here has been to clarify the main issues.

THE ROLE OF PHYSICIANS

If euthanasia is made legally permissible, should physicians take part in it? Should only physicians be permitted to perform it, as is the case in the Netherlands? In discussing whether euthanasia is incompatible with medicine's commitment to curing, caring for, and comforting patients, I argued that it is not at odds with a proper understanding of the aims of medicine, and so need not undermine patients' trust in their physicians. If that argument is correct, then physicians probably should not be prohibited, either by law or by professional norms, from taking part in a legally permissible practice of euthanasia (nor, of course, should they be compelled to do so if their personal or professional scruples forbid it). Most physicians in the Netherlands appear not to understand euthanasia to be incompatible with their professional commitments.

Sometimes patients who would be able to end their lives on their own nevertheless seek the assistance of physicians. Physician involvement in such cases may have important benefits to patients and others beyond simply assuring the use of effective means. Historically, in the United States suicide has carried a strong negative stigma that many today believe unwarranted. Seeking a physician's assistance, or what can almost seem a physician's blessing, may be a way of trying to remove that stigma and show others that the decision for suicide was made with due seriousness and was justified under the circumstances. The physician's involvement provides a kind of social approval, or more accurately helps counter what would otherwise be unwarranted social disapproval.

There are also at least two reasons for restricting the practice of euthanasia to physicians only. First, physicians would inevitably be involved in some of the important procedural safeguards necessary to a defensible practice, such as seeing to it that the patient is well-informed about his or her condition, prognosis, and possible treatments, and ensuring that all reasonable means have been taken to improve the quality of the patient's life. Second, and probably more important, one necessary protection against abuse of the practice is to limit the persons given authority to perform it, so that they can be held accountable for their exercise of that authority. Physicians, whose training and professional norms give some assurance that they would perform euthanasia responsibly, are an appropriate group of persons to whom the practice may be restricted.

NOTES

1 President's Commission for the Study of Ethical Problems in Medicine and Biomedical and Behavioral Research, *Deciding to Forego Life-Sustaining Treatment* (Washington, D.C.: U.S. Government Printing Office, 1983); The Hastings Center, *Guidelines on the Termination of Life-Sustaining Treatment and Care of the Dying* (Bloomington: Indiana University Press, 1987); *Current Opinions of the Council on Ethical and Judicial Affairs of the American Medical Association – 1989: Withholding or Withdrawing Life-Prolonging Treatment* (Chicago: American Medical Association, 1989); George Annas and Leonard Glantz, "The Right of El-

derly Patients to Refuse Life-Sustaining Treatment," *Millbank Memorial Quarterly* 64, suppl. 2 (1986): 95–162; Robert F. Weir, *Abating Treatment with Critically Ill Patients* (New York: Oxford University Press, 1989); Sidney J. Wanzer et al., "The Physician's Responsibility toward Hopelessly Ill Patients," *NEJM* 310 (1984): 955–59.

2 M.A.M. de Wachter, "Active Euthanasia in the Netherlands," *JAMA* 262, no. 23 (1989): 3315–19.

3 Anonymous, "It's Over, Debbie," *JAMA* 259 (1988): 272; Timothy E. Quill, "Death and Dignity," *NEJM* 322 (1990): 1881–83.

4 Wanzer et al., "The Physician's Responsibility toward Hopelessly Ill Patients: A Second Look," *NEJM* 320 (1989): 844–49.

5 Willard Gaylin, Leon R. Kass, Edmund D. Pellegrino, and Mark Siegler, "Doctors Must Not Kill," *JAMA* 259 (1988): 2139–40.

6 Bonnie Steinbock, ed., *Killing and Allowing to Die* (Englewood Cliffs, N.J.: Prentice Hall, 1980).

7 See Chapter 7 of this volume, "Forgoing Life-Sustaining Food and Water: Is it Killing?"

8 James Rachels, "Active and Passive Euthanasia," *NEJM* 292 (1975): 78–80; Michael Tooley, *Abortion and Infanticide* (Oxford: Oxford University Press, 1983). In Chapter 5 of this volume, "Taking Human Life," I argue in more detail that killing in itself is not morally different from allowing to die and defend the strategy of argument employed in this and the succeeding two paragraphs in the text.

9 See Chapter 4 of this volume, "Moral Rights and Permissible Killing."

10 P. Painton and E. Taylor, "Love or Let Die," *Time*, 19 March 1990, pp. 62–71; *Boston Globe*/Harvard University Poll, *Boston Globe*, 3 November 1991.

11 James Rachels, *The End of Life* (Oxford: Oxford University Press, 1986).

12 Marcia Angell, "The Quality of Mercy," *NEJM* 306 (1982): 98–99; M. Donovan, P. Dillon, and L. Mcguire, "Incidence and Characteristics of Pain in a Sample of Medical-Surgical Inpatients," *Pain* 30 (1987): 69–78.

13 Eric Cassell, *The Nature of Suffering and the Goals of Medicine* (New York: Oxford University Press, 1991).

14 Gaylin et al., "Doctors Must Not Kill."

15 Leon R. Kass, "Neither for Love Nor Money: Why Doctors Must Not Kill," *The Public Interest* 94 (1989): 25–46; cf. also his *Toward a More Natural Science: Biology and Human Affairs* (New York: The Free Press, 1985), chs. 6–9.

16 Paul J. Van der Maas et al., "Euthanasia and Other Medical Decisions Concerning the End of Life," *Lancet* 338 (1991): 669–74.

17 Susan M. Wolf, "Holding the Line on Euthanasia," Special Supplement, *Hastings Center Report* 19, no. 1 (1989): 13–15.

18 My formulation of this argument derives from David Velleman's statement of it in his commentary on an earlier version of this paper delivered at the American Philosophical Association Central Division meetings; a similar point was made to me by Elisha Milgram in discussion on another occasion. For more general development of the point see Thomas Schelling, *The Strategy of Conflict* (Cambridge, Mass.: Harvard University Press, 1960); and Gerald Dworkin, "Is More Choice Better Than Less?" in *The Theory and Practice of Autonomy* (Cambridge: Cambridge University Press, 1988).

19 Frederick Schauer, "Slippery Slopes," *Harvard Law Review* 99 (1985): 361–83; Wibren van der Burg, "The Slippery Slope Argument," *Ethics* 102 (October 1991): 42–65.

20 There is evidence that physicians commonly fail to diagnose depression. See Robert I. Misbin, "Physicians Aid in Dying," *NEJM* 325 (1991): 1304–7.

21 Richard Fenigsen, "A Case against Dutch Euthanasia," Special Supplement, *Hastings Center Report* 19, no. 1 (1989): 22–30.

22 Allen E. Buchanan and Dan W. Brock, *Deciding for Others: The Ethics of Surrogate Decisionmaking* (Cambridge: Cambridge University Press, 1989).

23 Van der Maas et al., "Euthanasia and Other Medical Decisions."

24 Margaret P. Battin, "Seven Caveats Concerning the Discussion of Euthanasia in Holland," *American Philosophical Association Newsletter on Philosophy and Medicine* 89, no. 2 (1990).

Part III

Life-and-death decisions in health policy

Chapter 9

The value of prolonging human life

How is the value of prolonging human life properly determined? This is a question for both consequentialist and non-consequentialist moral theories. Consequentialists quite obviously need an answer to it, since their program of acting so as to maximize value or the good requires an answer to how much value or good is produced by prolonging life. Virtually all non-consequentialists also grant that prolonging human life is usually of value or a good. When non-consequentialist moral constraints (whether of duty, rights, or justice) do not otherwise limit action or policy, non-consequentialists often seek to use efforts and resources to do the most good, and then need an account of the relative value of prolonging life.

The value of prolonging human life is, of course, not merely a problem for moral theory, but has great practical and policy importance in several areas: first, in the evaluation of a wide array of social, usually governmental, programs of environmental and safety protections and regulations whose explicit purpose is to save lives or to decrease persons' risk of death; second, in decisions about the acceptable level of risk to life that programs with other purposes may carry; third, in decisions about appropriate amounts to spend for life-sustaining measures such as health care. Perhaps because prolonging life has this policy importance in decisions about expenditures, both the theoretical and practical literature on principles and procedures for valuing lives has been dominated by economists and policy analysts. Philosophers, surprisingly, have had little to say on the issues. In order to try to bring the issues of concern for moral philosophy more closely to bear on the economic and policy treatments, my discussion is organized as an examination of the main alternative theoretical measures for valuing prolonging life that economists have proposed, assessing their underlying moral principles, assumptions and differences. One of my claims will be that both theoretical and practical treatments of the value of prolonging life need to distinguish more clearly than is often done between 'value' as moral value employed within moral theory and judgment, and value as economic or monetary value.

Finally, I want at the outset to emphasize the limits of the scope of my discussion. The theoretical measures I examine below are all methods for determining the *value* of prolonging *a* human life. Even a comprehensive treatment only of the *value* of saving human life would have to address many issues that there is not space to consider here, among them: statistical and identified lives; the symbolic importance of saving lives; empirical and rational attitudes towards risk. Moral evaluation of *actions* or *policies* prolonging life also raises issues about various non-consequentialist constraints on how the good of prolonging life is brought about, issues of justice about the distribution of life-prolonging efforts, and the moral relevance of consent to life-threatening risks. While I note a few places where these other issues arise, they require a separate treatment in their own right, which I do not attempt to provide in this essay.

PRINCIPLES FOR VALUING PROLONGING LIVES

Human capital and lost earnings

The human capital approach can be seen as an attempt to provide a direct measure of the economic gain or value of prolonging life, or loss in failing to do so. It perhaps deserves mention because it has probably been the dominant approach among economists and policy analysts until recently, but can be treated very briefly because it has been effectively criticized by others.[1] In this approach, the value of a life either saved or lost is measured by the future earnings stream saved or lost to the person.[2] More specifically, the value of a life lost is the current value of the expected future earnings lost to that individual by his early death, where current value is determined by applying a discount rate to future earnings adjusted for inflation. As Acton has remarked, this is a method borrowed from, and appropriate to, the evaluation of machinery and other goods used in economic production. It is essentially the method that is used in measuring the value of a capital good investment, transferred to investment in saving lives, and is inadequate just because the overall value of the loss of a person's life either to the person himself or to others is not equivalent to how much income he will produce. This might be an appropriate measure of the effect of the loss on GNP or National Income, but it seems quite to miss the basis of people's concern with saving lives, both their own and others.

Beyond the virtue of admitting of relatively precise numerical measure, there is little to be said for the human capital approach. It fails to capture or measure the loss that other persons who care about an individual would experience as a result of his death; for example, the value a parent places on the life of a loved child has little if anything to do with the child's expected income as an adult. And, more importantly, it fails to capture and measure why and how much an individual values his own life being prolonged. An adequate measure should reflect that an individual's concern not to be killed or allowed to die is not for the loss of his expected future earnings, but for the loss of his expected future experiences that he values and desires to undergo; to adapt the stream metaphor, it is his future experience stream, not earnings stream, that is relevant. Some of these experiences depend on his future earnings, of course, but many do not. Moreover, some people's desire to continue to live is not even based solely on their evaluation of the content and quality of the experience stream that will likely fill their lives.

They may desire continued life, or conscious life, in part for itself, whatever that life is like.

The lost earnings or human capital measure also creates several morally problematic differentiations in value between lives that an adequate measure should either avoid or provide explicit justification for. It makes the life of a high income person worth more than a low income person's, directly proportionate to the difference in their incomes. It discriminates against the elderly who, when retired, are no longer productive and no longer have earned income, as well as against disabled and retarded persons unable to work and whose death would produce no loss in earned income. This discrimination takes a strong form – if a person has ceased to be economically productive, saving his life has *no* value. If such persons' continued consumption is drawn from the earned income of others, their continued life would on this measure seem to constitute a net *dis*value. The human capital approach also discriminates against persons doing unreimbursed work or services, most importantly, women doing unpaid work in the home.

While the human capital approach measuring lost earnings continues to be employed in some policy analyses of the value of life-prolonging programs, it has come to be recognized by most as lacking any plausible moral basis as a measure either of the relative amount of good produced by prolonging a person's life, or of what it would be reasonable for that person or society generally to be willing to pay to do so. Perhaps the most widely accepted approach now, though difficulties of application have limited the empirical studies employing it, is the willingness to pay approach.

Willingness to pay

Just as philosophers have used the concept of utility, or an analogous concept, to provide a single measure of the good and bad effects of actions, so economists employing the willingness to pay test have used money as a single common measure of the costs and benefits of an action or policy. Some common measure is necessary in order to be able to evaluate whether the benefits of prolonging a life are worth the costs of doing so in comparison with other possible uses of resources. The basic idea in the willingness to pay approach is that the value of prolonging a person's life is determined by the value he or she (and others) in fact places on doing so, as in turn measured by what he or she is willing to pay to reduce a risk to life or to accept as compensation for an increased risk of death.[3] (I assume throughout that the willingness to pay test is applied in ideal conditions for judging, though I do not attempt to specify what those conditions are because they do not affect my concerns with willingness to pay.)

To illustrate, suppose the risk of death from myocardial infarction during the next year for a man of age 50 is 1 in 200, and that providing mobile coronary care units reduces that risk by 1%. Thus, this risk of death is reduced from .005 to .00495, or from 500 to 495 in 100,000. The person aged 50 is asked what he would be willing to pay for this reduction in risk of death from heart attack. If he were willing to pay, for example, $20 per year to support the mobile coronary care unit that reduces this probability of death by .00005, that is its monetary value to him and we can calculate that he values his life at $400,000. We could then perform similar inquiries with each of the other, say 9,999, persons who would be served by the unit, adjusting for differences in their age and risk of death, and conclude that if on average they too would each pay $20 the total monetary value to the individuals served of the benefits produced by the program is $200,000.

Many have noted that if we use a compensation (as opposed to willingness to pay) test for certain and immediate death, the test seems to fail. No amount of money would be enough to induce a person to accept it as compensation in exchange for certain and immediate death; it would then follow by this measure that an infinite value should be assigned to life.[4] (Unlike the willingness to pay version, which is limited by one's resources even for avoiding certain and immediate death, there is no limit to the possible compensation to accept such a death that an individual could ask.) However, this will only be so if we assume no bequests are possible and that the person values nothing else that the money might buy more than death. Also, if the condition that death will be immediate is relaxed, then a person might well accept a certain death at some future time which shortens his life expectancy, but greatly improves the quality of his life until his death, for example, by permitting some longed for and otherwise unattainable experience or accomplishment. The problem when death is immediate, certain, and bequests are barred is not that life is then of infinite value, but that money has no use to a dead person and so is then an inadequate measure of life's value.[5] Nevertheless, as a result of this difficulty the willingness to pay measure in its compensation version is usually restricted to increases or decreases in risks of death. (There is, however, an important difference in the value to an individual of avoiding risks of death as opposed to certain death, which the willingness to pay approach brings out and which I will explore later.)

There are several fundamental attractions of the willingness to pay approach to the value of life-prolonging programs. For economists, it conforms to basic assumptions of welfare economics. It satisfies their principle of "consumer sovereignty" according to which individuals are assumed to be, at least when choosing in ideal conditions, ultimately the best judges of their own welfare. The relative value of any benefit to a person is ultimately determined by the person's own judgment of its relative worth to him or her, and this is measured by the resources that the person will expend to obtain it. For philosophers, this feature conforms to any philosophical desire or preference theory of the good for persons. On any theory of human welfare according to which what serves a person's welfare is ultimately determined by his or her preferences (for example, his or her fully informed preferences made in ideal conditions for judging), willingness to pay will correctly measure the relative contribution to a person's good or welfare of reducing a risk to his or her life. As the example above illustrated, willingness to pay evaluations of overall life-saving programs, as a form of cost-benefit analysis, can be used to insure that programs are recommended only if they satisfy the potential Pareto principle, that is only if total benefits to all affected exceed total costs so that gainers from the program could compensate any losers and there still be some positive benefit. If methods can be devised for making such compensation, or for charging individuals according to willingness to pay (setting aside here public goods problems), recommended programs will be actual, not merely potential, Pareto improvements.

The method is also sensitive to different attitudes to risk among persons, as well as differences between persons in valuations placed on extension of life or prevention of early death. And, most importantly, it allows individuals to make their own trade-offs between the length of life and the quality of life; resources used to prevent or lower risks of death will be unavailable for use in increasing the quality of experience in life. A variety of differences between persons – in tastes, in the schedule of projects in their life plans, in opportunities and skills, and so forth – will result in different persons making different trade-offs between length and quality of life.[6] Willingness to pay thus respects one form of individual self-determination or autonomy by respecting persons' competent choices in the definition and pursuit of their own particular conception of the good or plan of life.

Finally, to the extent persons' attitudes to risks of death are affected by the kind of risk and not just its probability of death, this too will be accommodated. People do care about how they die, as well as when they die, and about the *kinds* of risks to which they are exposed. If death by cancer is feared to a degree inordinate to its probability in comparison with, say, death by automobile accident, a person will pay more, other things equal, to reduce the cancer risk.[7] The willingness to pay test is in principle sensitive to these differences, as an adequate test should be.

There are many obvious and well-known difficulties in the

practical application of willingness to pay when employed in empirical studies with real persons: for example, how well can persons reason about very small probabilities, how well informed are they about alternatives, how to avoid distortions from "framing effects," and so forth. I do not explore any of these here because my concern is with the explicitly moral issues for valuing prolonging life raised by the willingness to pay approach, even if the many practical difficulties have been overcome. The most obvious and common moral criticism of the willingness to pay approach is that it makes the value of prolonging a person's life depend in part on his or her income and wealth. Other things equal, the poorer a person is, the higher utility expenditures must be traded off to purchase a reduction in risk of death. Persons will not worry much about small risks of carcinogens or nuclear radiation until they have basic subsistence and the capacity to pursue their more important ends secured; if mobile coronary care units require trading food without which you will be hungry or malnourished, you may well prefer the food, but not if you must only trade down to a somewhat lower-quality growth Bordeaux to drink with your food. The willingness to pay approach seems to favor directing greater life-saving efforts at the rich, for example, putting mobile coronary care units in wealthy suburbs that would not be "worth it" in poor inner-city areas.

It is worth distinguishing two different versions of the objection to willingness to pay that it makes the value of prolonging or saving a person's life depend on the distribution of income. In the first version, the objection is that when the existing distribution of income is unjust determining the value of prolonging a person's life by his or her willingness to pay will only replicate this injustice, undervaluing saving the lives of those whose incomes are unjustly low, overvaluing those whose incomes are too high. In the second version, the objection is that even if one corrects the willingness to pay estimate for the unjust income distribution, and determines willingness to pay from a just distribution, nevertheless so long as the just distribution contains any income inequalities, as on most views it would, then the value of saving the rich will unjustifiably be greater than saving the poor. Put more concretely, life-saving measures like heart transplants or mobile coronary care units may turn out to be worth their cost for the rich but not for the poor, and this, so the objection goes, is morally unacceptable.

How serious is this objection, and exactly what does it come to? In what sense, if any, does willingness to pay commit us to valuing the lives of the rich more than the poor? In one interpretation, this objection is not specifically about the measure of the *value* of a life, but about the *justice* of distributing life-saving resources according to individuals' willingness to pay for them. There is a plausible case that it need not be *unjust* for the rich to have more life-saving resources and efforts than the poor, so long as income inequalities between rich and poor are not unjust. If money is an all-purpose means of access to other (though, of course, not *all* other) goods, and at least some income differences are assumed to be just, then why must any inequalities in goods for which the money is used necessarily be unjust? Even if it were assumed that *no* income inequalities are just and that persons' incomes were somehow made equal, differences between persons in preferences for prolonging life, or reducing risks (or particular kinds of risks) to life, as opposed to preferences for other goods, would still lead to differences between persons in resources spent on prolonging life. An argument from justice is needed that is sufficiently strong to require limiting people's freedom to use their justly acquired resources, whether unequal or equal, to prolong their own lives in ways that cause no harm to others. If some persons so value long life that they choose to deny themselves things others have in order to spend an unusually high percentage of their resources on reducing risks to their lives and providing for life-prolonging resources, why is it just to deny them the freedom to do so while allowing them to spend more than others on relatively frivolous goods like

243

Caribbean vacations and luxury cars? Doing so would be a deep infringement of people's self-determination or autonomy in forming and pursuing a life plan.

There could, however, be reasons of justice for not letting the distribution of survival probabilities be determined in the marketplace, whether or not the distribution of income is believed to be just. For varying reasons, the distribution of many goods, such as education, votes, rights to due process, and so forth, are kept out of the market, either in whole or in part. There might be good reasons to do so as well with survival probabilities. However, these distributive questions require a separate treatment; in this essay, my concern is with the determination of the value, not the just distribution, of prolonging life.

To many, willingness to pay remains morally offensive for its seeming implication that the lives of the rich are worth more than those of the poor. In a defense of the willingness to pay approach, Paul Menzel has sought to avoid that implication.[8] He argues that willingness to pay implies only that, *to the poor person,* prolonging his life at a particular cost will be worth less in comparison with other things he could buy with that money than the same prolongation would be worth, *to a rich person,* in comparison with other things he must forgo to purchase it. But this does not imply, Menzel argues, that there is any absolute sense in which the rich person's life, or saving the rich person's life, is more valuable than the poor person's. Each may care equally about his life and find it equally hard to give it up. Nor, Menzel adds, using an example from Schelling,[9] would it follow that a hospital that can save either of two persons, but not both, has any reason to save the richer or could claim that it would maximize welfare to do so. To think it would confuses the monetary value of the care to the individual involved with the overall value in some absolute sense. But the willingness to pay approach makes no use of a sense of absolute, overall value.[10] Does this remove the offensive implication?

Suppose the hospital has two patients, one rich and one poor, in similar medical need of the intensive care unit, but only one bed left in the ICU. Assume the income inequality is not unjust. Who should go in the more expensive ICU, and who should stay in the less expensive general hospital facility? For the reasons already discussed, doesn't the willingness to pay approach say the rich person should go in? If the hospital were to auction the bed to the two, the rich person, other things equal, would buy it. "Everything possible" and "decent minimum" health insurance policies might differ in that only the former funded ICU care in these circumstances. If the rich person had purchased, for reasons the willingness to pay approach explicates, "everything possible" coverage, while the poor person had bought "decent minimum" coverage, the rich person would be entitled to the care. Now the idea of a hospital or physician auctioning off scarce life-saving services to the highest bidder may seem morally grotesque. But I believe, contrary to Menzel, that the willingness to pay approach in essence supports such allocation devices as long as the underlying income distribution is not unjust. Except for the urgency of need for care, the auction is not in principle different from insurance, and willingness to pay proponents do support insurance as an allocational device to determine entitlements to health care. Thus, the willingness to pay approach does support procedures that allocate more, or scarce, life-prolonging resources to the rich than to the poor, though not because it assumes the lives of the rich to be of greater absolute value. The life-prolonging care or risk reduction has more economic or monetary value to the rich than the poor, but I want to argue that there is a philosophical sense of value, relevant to moral judgments about relative good or benefit produced, in which prolonging the life of each does not vary in this way with their wealth. In contrast to the monetary value of prolonging life delineated by willingness to pay, I shall call this latter the moral value of prolonging life.

To bring the difference out, suppose a physician grants that the economic or monetary value to the rich person of the ICU bed is greater, but wants to know simply whether he would do any more good by saving the rich instead of the poor person, or by providing him with care that gives him a higher probability of survival than the poor person. There is a sense in which it is reasonable to assume that the value or degree of benefit to each would be the same, or at least does not vary with wealth and ability to pay, though the monetary value of the benefit to each differs. Compare the case of alleviating pain. Many believe that pain is bad or evil, no matter whose pain it is.[11] To prevent or remove another's pain is to do him good or to benefit him. How much the good or benefit is depends on how great the pain is, or on how much suffering it causes, but not on how rich the sufferer is or on how much he or she would pay to have the pain alleviated. We can and do distinguish the size or degree of good or benefit conferred, from how much its recipient would be willing to pay to get it; only the latter varies with the wealth of the recipient. Why isn't prolonging human life for a given amount of time, or reducing a given risk to life, also a good or benefit of the same amount for anyone, or at least a good or benefit that does not vary with the beneficiary's wealth and consequent willingness to pay for it? Distinguishing the amount of good or benefit in this way from its different monetary value to different recipients makes sense of the idea that the rich can use their wealth to obtain more good or benefit for themselves. So the doctor (or policy analyst) who wants to do the most good has no reason to distinguish between prolonging the lives of the rich or the poor; other things equal, the moral value or benefit of prolonging a person's life is the same in each case. But then what is the relation between the willingness to pay approach and the moral value or good from prolonging different persons' lives?

It would be a mistake to hold that the degree or amount of benefit of prolonging life of a specified quality and duration does not vary at all according to the preferences of the subject. An account either of the economic *or* of the moral value of prolonging life should reflect the fact that different persons reasonably assign greater or lesser value to having their life prolonged. Because of differences in persons' overall life plans, length of life, like other goods such as health, may for some but not others reasonably be sacrificed in order to secure other important ends. A measure of value is needed that allows for differences between persons based on differences in their settled preferences, but that does not differentially weight those differences according to differences between persons in income and wealth. There are a number of ways such a value scale might be developed and formidable, though not in my view insuperable, difficulties in doing so. Many of the familiar problems in making interpersonal comparisons of utilities arise, and I shall not enter into any of their details here. Very roughly and intuitively, one alternative is to assume persons have the same overall total units of value to be differentially distributed by each in a given period to particular alternatives (including prolonging life) according to their different preferences. This essentially assumes equality in the total weight given to the preferences of each person for a given period. In this account, the good produced by prolonging the lives of the rich is not, other things equal, more than what is produced by doing so for the poor, though it would still be true that the rich would pay more to prolong their lives or to reduce risks to their lives. Willingness to pay then does *not* measure the amount of the benefit or good done, what I have called the moral value, for a person by reducing a risk to his life; the willingness to pay to reduce the risk *does* vary with income, whereas the benefit or good for the person in an interpersonal sense of benefit or good does not.

But it is clearly an interpersonal sense of good, benefit or moral value that moral philosophers require for any consequentialist or non-consequentialist moral theory in which judgments of overall benefits or good requiring interpersonal trade-offs are made. Economists, on the other hand, commonly assume that interpersonal utility or welfare judgments are impossible, or at least problematic. Willingness to pay tests, summed over all affected by a program, then have the attraction of requiring appeal only to a principle of Pareto Optimality which does not require interpersonal comparisons. Given any particular individual's income, the willingness to pay criteria will secure for him the amount of life prolongation or risk reduction as opposed to other goods that will maximize his overall welfare, and no interpersonal welfare judgments are necessary. In a nutshell, willingness to pay tests correctly ascribe different economic or monetary value to reducing risks to life to rich and poor, while the amount of good or benefit relevant to the moral evaluation of consequences or states of affairs should not so vary.

Whether the willingness to pay test is appropriate for determining the value of prolonging life or reducing risks to life thus depends on distinguishing exactly what question is being asked more clearly than is often done. Willingness to pay *is* appropriate for questions such as the following. If the distribution of income is simply taken as given, how much should each individual spend on prolonging his or her life as opposed to other goods in order to maximize his or her welfare? Likewise, if the distribution of income is independently determined to be fair or just, how much should be devoted by each individual toward prolonging his life in order to maximize his or her welfare? How much should society or government devote from a given amount of resources (whose distribution to different individuals or groups is fixed) to prolonging the lives of those individuals or groups as opposed to promoting their other ends in order to maximize their welfare? By providing a test for how much resources each individual should devote to prolonging his or her life, and assuming no interpersonal welfare judgments and no income redistribution, willingness to pay tests can determine how much any group of individuals should devote to prolonging life to maximize the welfare of all members of that group. On the other hand, if interpersonal welfare or utility comparisons are possible, willingness to pay tests will *not* answer questions such as the following. When and how should income be redistributed so that willingness to pay tests applied to individuals will determine the spending for prolonging life for those individuals that will maximize welfare or utility (in the interpersonal sense) for the group? How

should social or governmental resources be apportioned between prolonging life and other ends for particular individuals so as to maximize welfare or moral value (in the
interpersonal sense) for all affected? These are, of course, by
no means all of the questions in which the value of prolonging life arises. I use them only to illustrate the necessity
of distinguishing whether what is in question is monetary
value or what I have called moral value in an interpersonal
interpretation.

With this distinction between the moral value or benefit
provided an individual and the differential monetary value
of that benefit to the individual depending on his income
and what else he could buy instead, we are in a position to
bring out an important difference in whether one is facing
certain death or only a risk of death, if no steps to avoid the
threat are taken. Policy-makers selecting programs for funding might well decide on the basis of willingness to pay
estimates that a particular safety program reducing risks to
life was worth its economic cost to relatively well-off groups,
but not to the poor. For the poor, greater benefits would be
obtained if the resources were used on programs providing
benefits to which the poor gave higher priority. Now suppose
the policy-maker is a physician or health care administrator
considering how a scarce life-saving medical treatment is to
be distributed when more need it than will be able to get it.
Without the treatment, those in need of it will with certainty
now die. Should the rich be favored over the poor in order
to produce the most good with the resources, as in the case
above of risk? Assuming that rich and poor differ only in
their economic situation and have the same relative preferences for prolongation of their lives, I have already argued
above that on a plausible account of interpersonal moral
value the benefit to each of prolonging their lives is the same.
No more good is done by treating one or the other. Unlike
the case of risk where death was uncertain, there is no
alternative of maximizing the good by giving the life-
prolonging treatment to the rich person and giving the poor
person the money to use on some higher-ranked unsatisfied

preference instead. Without the life-sustaining treatment, he will be dead and so be unable to use the money to satisfy any higher-ranked preference. A different way of putting the point is that if faced with *certain* death without life-sustaining treatment, getting that treatment will then become the poor person's highest-ranked preference and so there will be no better uses to which he might put the funds used for life-sustaining treatment, as was the case with only a *risk* of death. In this instance the *monetary value* of life-saving will be greater to the rich person *only* because he has more money available; faced with certain death without treatment, it is rational to use all his resources to secure treatment and avoid death. It will not be greater because, as in the case of risk reduction, the poor person has a higher-priority, higher-valued end than the risk reduction – when loss of life is certain, he will have no higher-priority alternative use of the money. Consequently, in distributing life-saving resources necessary to avoid certain death for known individuals, there is no basis for distinguishing between rich and poor in maximizing the good produced by the resources – if rich and poor are otherwise similar, it is indifferent to whom the scarce resources go.[12]

One implication of this difference between the cases of certainty and uncertainty is that, if we seek to maximize the good in the use of resources, it is of crucial importance when and under what conditions decision-making takes place. At the point of risk reduction under uncertainty, more good may be produced by risk reduction measures for the rich and alternative higher-priority measures such as improved housing for the poor. On the other hand, at the point where uncertainty concerning risks has been removed and both rich and poor face certain death without life-sustaining measures, no more good is then done by saving one than the other simply because of the economic difference between them. The example of life-sustaining health care, which is generally secured at the point of uncertainty through insurance, illustrates the difficulty this poses. Faced with uncertainty about future needs for expensive life-sustaining medical care, a rich person may buy "everything possible" health insurance while a poor person reasonably chooses a "decent minimum" policy, using the difference in cost for other higher-priority needs. Now suppose both later come to need an organ transplant whose cost is covered by the everything possible policy, but not by the decent minimum policy. The rich person but not the poor person will then be entitled to the life-saving care as a result of their earlier decisions that, for each, produced the most good possible, given their different incomes. At the later time of need, one is entitled to the care, the other is not, while there is no difference in the amount of the benefit to each; each will suffer as great a loss as the other without the transplant. Their earlier insurance decisions in conditions of uncertainty give reason to discriminate between the two now, though the same value is produced or loss prevented for each if they are instead treated the same. There is nothing inconsistent or paradoxical in this difference in what does the most good at different decision points with life-prolonging resources; analogous disparities are familiar elsewhere. But it does underline the moral importance of when the decision is effectively made about the use of life-sustaining resources if we seek to maximize the benefits of resource use. And it illuminates as well why, though the earlier choices of different insurance packages may have been both just and rational if the income difference was just, differentiation between rich and poor in distribution of this most important benefit, at the point where the loss to each without it is the same and of "everything" is often seen to be morally problematic.

Quality-adjusted life years

A third approach to measuring the value of prolonging life is commonly employed in cost-effectiveness analyses of life-saving programs. It does not place a dollar value on a life saved, but instead measures the cost of a life saved, a life year saved, or a quality-adjusted life year saved in a given program. Whichever of these three units is employed, alternative programs can then be compared for their relative cost effectiveness in saving lives. Unlike willingness to pay, this approach gives no account of how much resources should be devoted to prolonging lives or reducing risks to life as opposed to other ends. But given a fixed amount of money to spend, or programs of a fixed nature, it provides a comparison of their relative effectiveness in saving lives. It tells us when programs save lives at very high cost, or when they are cheap, or perhaps even free because of other side benefits.[13] Which is the appropriate goal to be measured – lives, life years or quality-adjusted life years?[14] Unlike willingness to pay, which uses the relative preferences of persons to establish the relative value of prolonging life, cost-effectiveness analyses seek to isolate a physical unit of measure which is the goal valued by persons in life-prolonging programs. With important qualifications to be noted, I shall argue that Zeckhauser and others are correct that the measure of value in cost-effectiveness analyses of life prolongation should be the quality-adjusted life years measure (QALYs), though we will see here one plausible distributive constraint on any maximizing policy regarding QALYs.[15]

What exactly is the cost per life saved measure? Of course, nothing anyone does can save a life permanently, since it is not possible to make anyone immortal; the most that can be done is to prolong life and postpone death. Any measure of lives saved then presumably measures a reduction in the loss of life during a given period from a particular cause as the result of some change introduced, for example, a reduction in deaths during a year from myocardial infarction as a result of the introduction of mobile coronary care units. Is the value of saving lives invariant with respect to the life expectancy of those whose deaths were prevented? At least other things equal, individuals generally are not indifferent between alternatives that prolong their lives for different lengths of time. Willingness to pay proponents point out that a person will be prepared, other things equal, to pay more for a year's prolongation of life than a week's assuming only a constant positive assessment of life during that period.[16] This reflects that the individual concerned values the longer extension more, and takes it to be a bigger benefit. If instead of the preferences or valuings of the individual in question, we look to the experiences that would fill the two periods, then whether they are valued for their pleasure, happiness, satisfaction of desire, or being experiences of specified valuable sorts, in each case, other things equal, the longer the period the greater the quantity of those valuable experiences, and so the greater the value.

Does this show (leaving open the possibility to be taken up below of a quality adjustment) that the longer the prolongation of life the better, and that the value produced increases in equal increments with each additional unit of life prolongation, as the life years approach suggests? I believe that is difficult to sort out and clarify, in part because many persons' judgments of the value, good or benefit produced are intermixed with complex and often unarticulated distributive considerations. Assuming no differences in quality, if a given individual's life is prolonged for two years instead of one, most would say that he has been done twice the, or at least a greater, benefit. In various fictional and philosophical accounts, the argument has been made that immortality would be tedious to the point where additional years would become no longer benefits, but burdens.[17] But this is a point, I believe, about what would happen to the quality of persons' lives if they extended indefinitely. If individuals were able sufficiently to adjust their life plans to great or indefinite extensions of life, and physical deterioration were arrested, it is unclear either that their quality of life *must* decline or that additional life years must lose value. Thus, other things equal, perhaps 10 additional years are ten times as good or valuable for a person as 1, and 100 ten times as good as 10. But even within a single life, where issues of distributive justice do not arise, these judgments are not entirely unproblematic.

Persons' life plans have certain common shapes and patterns reflecting the length of the normal human life span and changes in people's physical conditions and other concerns over that life span. We shape our plans, projects, activities and commitments to our reasonable expectations of the life span we will have. Thus, prolongation or continuance of life at different ages, though of equal duration, may be of quite different value and importance. Prolonging life 10 years for a person at age 50 who can then complete and carry out the projects, plans and commitments that have shaped her life may be vastly more important or valuable to that person than extending her life an additional 10 years at age 80 when the extended projects and commitments of her life plan are essentially completed, even assuming no significant physical or mental deterioration. I believe a sophisticated quality measure would take account of this differing value persons commonly assign to additional years at different periods in a life plan, just as it would take account of quality differences associated with the normal aging process. If so, perhaps within a life additional life years produce additional increments of value for a person, though the quality adjustment will then be more complex than it might at first have seemed.

When the life years saved accrue to different individuals, judgments of the good produced are complicated by distributive considerations. As I noted at the outset of the chapter, distributive issues require a separate treatment that I do not attempt here, but I do want to note briefly one particularly interesting distributive argument that illustrates their general importance. Suppose the normal life span for humans is around 70 years. Persons learn this as they grow to maturity and, barring special circumstances which alter that expectation in their own case, they plan their lives accordingly so that major plans and projects fit within these constraints. They also budget resources to attempt to maintain a reasonable level of happiness and well-being over that period, or at least to prevent their levels falling below some floor.

How should we morally evaluate life year extensions beyond the normal life span for some individuals or groups within a society, compared with prolonging the lives of others to bring them up to, or closer to, the normal life span?

Doing the latter not only brings them closer to the normal and expected quantity of experience, but can permit them to construct, integrate and complete a normal life plan. I will not detail the argument here, but I believe a plausible conception of equality of opportunity can be extended from its more familiar application to basic goods necessary to the choice among a normal array of life plans, such as education, to life years themselves.[18] The opportunity to choose a life plan from among an array normal in a particular society requires goods like education which develop skills and abilities necessary for pursuing such plans, but it even more obviously requires the life years to pursue a normal life plan. It is this idea, I believe, which underlies common feelings of unfairness (even if perhaps only by extension a sense of "cosmic unfairness," a sense that "life isn't fair," when it is nobody's doing or fault and nothing can be done about it) when an individual's life is cut off at an early age or in the prime of life; his life and its projects and goals were just getting underway or were cut off midstream in development. When early deaths can be prevented, deploying life-saving resources and efforts responds to a claim of equality of opportunity on the part of the person whose life is prematurely threatened.

This is similar to the claim of the handicapped for the extra compensatory efforts and resources necessary for them to pursue a normal life plan. Both claims are grounded in equality of opportunity, and do not presuppose that meeting them will maximize total welfare or value – either total life years or other satisfactions. The equal opportunity claim to the life years necessary for a normal life plan will not always be a decisive claim: the opportunities forgone to save others' lives or to produce other non-life-prolonging benefits may conflict with and sometimes override it. But there is a special moral priority in focusing life-prolonging efforts on individuals' opportunity to reach a normal life span instead of on extending lives beyond the normal life span, quite independent of any declining marginal quality of life years beyond the normal life span. Focusing on preventing early deaths not only produces more life years saved, but saves life years which have a greater moral claim on our efforts.[19]

How is the quality-adjusted life years approach different from the willingness to pay approach in the evaluations it yields of programs and efforts that save lives or reduce risks of death? Will deploying life-saving resources so as to maximize QALYs yield the same conclusions as deploying them on projects for which the summation of individuals' willingness to pay is greatest? One difference is obvious and important. As already noted willingness to pay approaches will be sensitive to differences between persons in income and wealth. Projects that affect high- as opposed to low-income groups will be favored disproportionate to their effects in the production of QALYs. There is a condition of equality of persons built into the QALY approach, and specifically an equalizing for differences in income and wealth, that is absent in the willingness to pay approach.[20] In this respect the QALY approach implicitly attends to the distinction noted in the willingness to pay section between the moral value of the good or benefit produced for a person if his life is saved or a risk to his life reduced and the monetary value of that benefit to him. The QALY approach measures the moral value to different individuals of saving their lives or reducing risks to their lives, which is invariant with respect to differences in their income and wealth. This is appropriate when the distribution is of a life-prolonging budget among different life-prolonging activities, since the value of the benefits of those activities will not vary with wealth or willingness to pay of the beneficiaries. No cost-effectiveness analysis, however, will tell us when a particular life-prolonging program is not worth its cost to low-income groups in comparison with other uses of those resources for them, while worth its cost to high income groups.

There is another possible difference of importance between QALY and willingness to pay, depending on how the latter measure is interpreted. Suppose, for genetic or other health reasons about which nothing can now be done, persons already have differences in life expectancies, so that A at 50 has a life expectancy of 5 years and B also at 50 has a life expectancy of 25 years. Assuming no differences in average quality and no discounting, reducing a given risk of death tomorrow for B instead of A produces 5 times the value under the QALY approach. However, assuming similar incomes, would B as a result be willing, all other things equal, to spend 5 times as much as A to reduce the risk? Each might be enjoying and valuing their life equally. B has five times as much life as A to look forward to, and so there is a sense in which B loses 5 times as great a good or benefit as A. This sense is measured by the QALY approach. However, there is another sense in which each loses everything if he dies tomorrow; each suffers the same loss of being reduced to a state of nonexistence.[21] Given similar incomes and preference orderings, it would seem that A might be willing to pay as much as B to avoid this loss. The issue of interpretation of willingness to pay, about which its proponents are generally unclear, is whether it should be determined out of one's present income and wealth, or out of one's present *and* total expected future income and wealth. Willingness to pay can probably be brought into line with QALY measures in cases of substantial differences in life expectancy if it is assumed that a person determines how much he is willing to pay from present *and expected future* earnings, for example, by permitting borrowing against future earnings to pay to reduce or eliminate a present risk. This will also permit extension of the willingness to pay approach to cases of infants and children who do not yet have an income, and will flatten out differences in what a particular individual at different times in his life would pay to reduce a given risk because of differences in his income at those different times. While the interpretation of willingness to pay from lifetime earnings is probably theoretically preferable, it clearly adds formidable difficulties to its practical application.

Both the willingness to pay and QALY approaches to valuing life-saving make the value vary according to the quality of life of the person saved. The QALY approach does so explicitly by adjusting life years saved for differences in the quality of those life years. As applied to large-scale social programs, the adjustments generally are only for significant group differences; for example, a lower quality is assigned to life years for the elderly, or for seriously disabled or handicapped groups, than to life years for normal adults. As applied to the use of scarce life-saving or risk-reduction resources for individual persons, the quality adjustment can be more fine-grained in taking account of more individual variations, including variations arising out of differences between persons in their preferences and values as well as differences in their objective circumstances and conditions. The QALY approach uses common quality adjustment criteria for different persons only as a crude, though useful, practical approximation. The willingness to pay approach adjusts the value of life to quality differences to the extent that individuals themselves adjust their willingness to pay for life extensions or risk reductions to differences in the quality of the additional life secured. And it seems clear that in general individuals in fact will do so. Individuals attempt to arrange their futures so as to secure a higher quality of life. It would seem uncontroversial that additional years of life with one's normal capacities intact are worth more than the same years with substantial disability from diseases like stroke or Alzheimer's. Individuals quite reasonably would be willing to pay more for health care insurance that funds expensive life-sustaining medical treatments like renal dialysis or organ transplants in circumstances when they are otherwise healthy than when they will be seriously incapacitated even with the treatment. In the extreme case, the quality adjustment can make the overall value of additional life negative, as when patients decline further life-sustaining treatment because they judge the life possible for them to be worse than no additional life at all. The details of how quality adjustments should be made, and how persons do in fact make them, are complex; for example, some serious physical disabilities may be fully compensated for by many persons so as to require no lower quality adjustment, and some disabilities, such as mental impairments that affect an individual's sense of personal identity, may require especially large downward quality adjustments. However, there is not space to enter into these details here.

I believe that both willingness to pay and QALY approaches are correct that the value, in both the moral and economic sense, of prolonging life varies according to the quality of the life prolonged. Objections to use of "quality of life considerations" in decisions about prolonging life on closer inspection often express a concern to protect particularly vulnerable populations and generally do not reduce to claims that the value of the benefit of prolonging life is invariant with respect to quality, though I shall not take the time here to show that. There is one point of some theoretical importance about the relevance of quality of life adjustments that should be noted. It turns once again on whether the decision about prolonging life will make death a certainty or only more likely, that is, whether it entails failing to act when death is otherwise certain or merely failing to reduce a risk of death. When evaluating how much to expend in order to reduce a risk to one's life or to insure against a possible need for life-sustaining measures, one can reasonably forgo reducing some risks of death in order to have the resources the risk reduction would require available for other quality of life enhancing uses in the meantime or if the risk does not eventuate. On the other hand, if it is a matter of whether to use resources to prolong life when death is otherwise immediate and certain, it is no longer reasonable to forgo the prolongation to save resources for some other use for that person.

The implication for quality of life considerations is that there is then only one significant quality of life distinction for an individual in that latter decision situation – will the additional life be of sufficient quality that it is preferable to no further life at all? So long as the prospective life passes that minimum quality threshold, further differences in quality are not relevant to what it is then reasonable for the person whose life it is to pay or forgo to secure his life. Whether the quality is of very low, but still positive, value, or of very high positive value, there is no reason to hold back any available resources for other uses if certain and immediate death (and again assuming no bequests) is the alternative prospect. It is only necessary for the quality of the prospective life to be of positive value for it to be worth choosing by the person in question, and worth using whatever resources at his disposal to secure. Other differences in quality will not affect his choice.

Both the willingness to pay and the QALY approaches to the value of saving lives, as I have characterized them to here, are methods of determining the value to an individual of the loss or saving of his or her life. Neither approach as so far characterized takes into account the value of an individual's life to others in order to arrive at a determination of its total monetary or moral value. In theory, if not in practice, the willingness to pay approach is relatively easily extended to include an individual's value to others. Others are simply asked what they would be willing to pay to reduce a risk to, or to save, the individual in question. Two studies have estimated an increase on average from relatives' willingness to pay of either 50% or 90% over the value to the individual.[22] This introduces an additional form of differentiation in the value of different lives from that based on income and wealth. Once we include value to others, prolonging the lives of persons who have a wide group of relations, loved ones, and friends, and on whom others importantly depend, will have comparatively greater value than will prolonging the lives of more socially isolated individuals without immediate families and friends who care about and depend on them.

Should this factor affect the monetary or moral value of prolonging particular individuals' lives? As many have noted, if the desires of others that a person be saved are to be counted towards the overall value of saving him, the desires of others that the person not be saved and instead die would seem to be relevant as well. These latter desires may express "sheer malevolence," but they may also merely reflect the bad effects for others if an individual's life is saved or prolonged, just as others' desires that a person be saved will often reflect the good effects of doing so for those others. Once again, whether the good for persons or moral value is construed in terms of pleasure, happiness, the satisfaction of desire, or the having of specified sorts of experiences, it seems clear that prolonging the life of one individual may often contribute to (or detract from) the good of others so construed. When that is the case, it would in turn seem that any account of the *overall* value of prolonging an individual's life must include the value or good produced for others, as well as for the individual, by doing so.

Menzel has argued that variations due to income differences in people's willingness to pay pricings of their lives are morally acceptable only if the poor can use their 'savings' from lower investments in risk reductions to meet their other needs.[23] They would be able to do so if they received cash instead of in-kind risk reduction, or if all had fair incomes from which they purchased risk reductions, with the poor then buying less than the rich in order to secure other higher-priority ends. Either scheme would increase the poor's welfare while respecting their autonomy to act on their own preferences, in comparison with forcing them to buy risk reduction in the higher amounts desired by the rich. Menzel suggests there is no appropriate recipient of the 'savings' that accrue from differential investments in risk reductions as a result of counting the preferences or willingness to pay of others. But that seems neither the case nor the issue. If the preferences of others for risk reduction to a particular individual are not counted, then welfare will not be maximized because the welfare increase to others from reducing risks to that individual will have been ignored, and the preferences of those others for reducing that individual's risks will not have been respected. In the scheme of fair-income shares from which risk reduction is purchased, others who value risk reduction for an individual would use some of their income to help purchase risk reduction for that individual. Persons who value others' risk reduction less in comparison with other goods because they care less about the survival of others would use their 'savings' on risk reduction to secure more of their other ends. This differential 'savings' because less is spent on risk reduction appropriately accrues not to the individual on whom less is spent, but to those who choose to spend disproportionately less than do others to reduce that individual's risks. If this is morally unjustified discrimination against the socially isolated, it will have to be for reasons other than whether saving them or reducing risks to them in fact generally does produce less value for others, and so, other things equal, less overall value. There is no theoretical reason to ignore the value or disvalue to others of prolonging an individual's life in determining the overall value of doing so.

Let me summarize very briefly, and omitting important qualifications, some main conclusions of my discussion of alternative measures employed in policy analyses of the value of prolonging life. The human capital approach lacks any plausible moral basis. Measuring the value of prolonging, or reducing risks to, an individual's life by his or her willingness to pay has more promise. Ultimately, it leaves individuals sovereign in the determination of the nature of their own good, including the importance of prolonging life in their good. It attends to differences between individuals in their preferences and values, especially differences in trade-offs between life extension and specific changes in their quality of life. However, in making the value of prolonging individuals' lives vary according to their income and wealth, it provides an economic measure of the monetary value of prolonging life to an individual. I have shown that this must be distinguished from what I have called the moral value to individuals of prolonging life, which does not vary with differences between persons in income and wealth. Neither the monetary nor moral value of prolonging life is always the correct interpretation. Rather, different questions about the value of prolonging life, and about how resources should be expended on prolonging life to maximize value, must be more carefully distinguished than is often done to be clear just which sense of value is appropriate.

I have suggested how willingness to pay will be determined quite differently depending on whether it is for a life-prolonging effort without which death will be immediate and certain or for a reduction in a risk to life. This difference has important policy implications that I have not fully explored here, for example, in thinking about spending for safety measures as opposed to life-prolonging medical treatment.

Unlike willingness to pay, the quality-adjusted life years approach does not employ a common unit of measure that permits a comparison of the value of prolonging life with other activities. While it ignores differences between persons in their ability to pay for life-saving, and so in this respect approximates the moral value of prolonging life, in most other respects it is similar to willingness to pay. The extent of the differences in the two approaches depends in part on details of their interpretation, particularly the interpretation of willingness to pay, on which their proponents are often unclear. I have suggested that where a particular life year extension falls within a normal life plan, and specifically whether the additional life year is within or beyond the normal life span, may affect its quality-adjusted value. I have also illustrated how distributive considerations are important to individuals' moral claims on efforts and resources for life year extensions with the suggestion that persons have a claim grounded in equality of opportunity to the life years necessary to a choice from among a normal array of life plans in their society.

Finally, I have argued that the value of extending life should vary both with the quality of the life extended, though I have not explored the complex nature and details of quality adjustments, and with the value to others of extending the individual's life. Both these features of the *value* of saving lives underline the importance of two further issues. One I have touched on but by no means fully explored – the relation of the value of extending life to a *just* distribution of survival probabilities. The other has not even been mentioned, but has nevertheless been lurking in the background of some of my discussion – the relation between prolonging or not prolonging life and *taking* life, and in what respects, if any, moral evaluations of taking life should also vary with such factors as the life expectancy, quality of life, and value to others of the victim.

NOTES

1 See, for example, E. J. Mishan: 1971, 'Evaluation of life and limb: A theoretical approach.' *Journal of Political Economy* 79, no. 4, pp. 687–706; and J. P. Acton: 1976, 'Measuring the monetary value of life-saving programs,' *Law and Contemporary Problems* 40, no. 4, pp. 46–72.

2 I will use "lives prolonged" and "lives saved" interchangeably, but prefer 'prolonged' because strictly that is the most we can ever do and because it focuses on the trade-off between longer life and other ends.

3 Standard expositions of the willingness to pay approach in the economics and policy literature can be found in: Mishan, *op. cit.*; Acton, *op. cit.*; Richard Zeckhauser: 1975, 'Procedures for valuing lives,' *Public Policy*, 23, no. 4, pp. 419–464; and M. W. Jones-Lee: 1976, *The Value of Life* (Chicago: University of Chicago Press).

4 John Broome uses this fact to mount a wholesale assault on willingness to pay measures of the value of a life in 'Trying to value a life,' *Journal of Public Economics* 9 (1978), pp. 91–100. Broome's argument is criticized by M. W. Jones-Lee: 1979, 'Trying to value a life,' *Journal of Public Economics* 12, pp. 249–256.

5 This inability to compensate for certain loss of life may seem a possible explanation for why killing some persons is commonly viewed as not justified even when necessary to save a greater number of others. The harm of death to those killed is certain and it is then difficult if not impossible to compensate them for the harm. But because certain death is one of the greatest harms, compensation for it is proportionately more morally urgent in the absence of otherwise showing that it is fair to sacrifice the one killed, or that he deserves to be sacrificed, for the benefit of others. In this account, the strength of the moral prohibition of killing, in comparison with other harms, derives in part from the extreme difficulty, and usual impossibility, of compensating the victim for certain death.

This cannot, however, explain a moral difference between killing and allowing to die since immediate and certain death from being allowed to die is equally difficult to compensate. Thus, something more is needed to explain why killing some to save more is wrong. I believe what is needed is to show how it can be fair to the ones killed to do so, but a comparable explanation is needed for those allowed to die in order to save a greater number of others.

6   It is undeniable that individuals in fact make decisions that either explicitly or implicitly trade quantity against quality of life. Nevertheless, some persons will retain a lingering sense of incommensurability between quantity and quality of life – what could be the common unit of measure by which they are valued? This raises general issues about the meaning and basis of claims for the incommensurability of values that I cannot take up here. For one possible account of the commensurability of values, see James Griffin: Fall 1977, 'Are there incommensurable values?' *Philosophy and Public Affairs* 7, no. 1, pp. 39–59. All that the willingness to pay approach requires, however, is that persons do in fact have ordered preferences for different quantity-quality packages.

7   There is an important and difficult problem lurking here of how to distinguish rational desires regarding how one dies from irrational attitudes towards particular risks that on at least some versions of willingness to pay should be discounted or disregarded.

8   Paul Menzel: 1983, *Medical Costs, Moral Choices* (Yale University Press, New Haven), Chs. 2, 3.

9   Thomas C. Schelling: 1984, 'The life you save may be your own,' in *Choice and Consequence* (Harvard University Press, Cambridge).

10 That there is any meaningful sense of absolute overall value of lives is also denied in Nicholas Rescher: 1983. *Risk: A Philosophical Introduction to the Theory of Risk Evaluation and Management* (University Press of America).

11 Cf. Thomas Nagel: 1970, *The Possibility of Altruism* (Oxford University Press, Oxford).

12 The argument of this paragraph assumes that in the case of an individual faced with certain death, bequests are precluded by which he could insure that after his death the funds will be used for some purpose he values. That is an artificial as-

sumption, and so even when faced with immediate and certain death in the absence of some life-sustaining measure, there often are alternative uses for funds on which the individual places some value. To that extent, such cases still allow trade-offs of the sort involved in willingness to pay decisions to reduce risks to death. But just because a person's other ends and values tend to pale in importance if he is faced with preventable, but otherwise immediate and certain death, the pure case without bequests is illuminating.

13   A survey of such cost-effectiveness evaluations of various federal programs shows the range in costs to be enormous: from zero cost per life or life year saved in a number of highway safety programs in which reduced costs of morbidity cover the program's full cost and gains in mortality come as an added dividend, to the most expensive program, the regulation of Acrylonitrile proposed by OSHA estimated to cost $11,000,000 per life year saved, and $169,200,000 per life saved. Cf. J. D. Graham and J. W. Vaupel: 1981, 'Value of life: What difference does it make.' *Risk Analysis* 1, pp. 89–95.

14   It is important to distinguish assigning value to saving the lives, life years, or quality-adjusted life years of existing persons from assigning value to additional lives, life years or quality-adjusted life years. The cost-effectiveness analyses of life-saving programs under consideration here employ a version of the former view. The latter view has important, and largely undesirable, implications for population policy which I shall ignore here.

15   Richard Zeckhauser, *op, cit.*

16   Though if death is otherwise certain and immediate and bequests are precluded, a person who might pay everything he has for a year's prolongation might do the same for a week's prolongation if that is the best he can do. Willingness to pay then would not yield a difference between the two, but it would not follow that the person was indifferent between the two alternatives. This is only a particular instance of the general failure of willingness to pay with cases of immediate and certain death when bequests are excluded.

17   One example is Bernard Williams: 1973, 'The Makropulos case: Reflections on the tedium of immortality,' *Problems of the Self* (Cambridge University Press, Cambridge).

18   Norman Daniels has applied a fair equality of opportunity

argument to health care generally in 'Health care needs and distributive justice,' *Philosophy and Public Affairs* 10, no. 2 (1981), pp. 146–179, and in his *Just Health Care* (Cambridge University Press, Cambridge, 1985). However, Daniels prefers a prudential insurance approach to the question of differences in an individual's claims on resources at different periods in his life; cf. his 'Am I my parents' keeper?' *Midwest Studies in Philosophy,* Vol. VII (1982), pp. 517–540.

19 For various arguments bearing on the importance of preventing early deaths, cf. James W. Vaupel: 1976, 'Early death: An American tragedy,' *Law and Contemporary Problems,* 40, no. 4, p. 73.

20 It overstates matters to say that the QALY approach *fully* equalizes for differences between persons in income and wealth. The QALY approach does measure only life years saved, adjusted for differences in their quality, without regard for whose life years they are or the economic position of the person whose they are. However, differences between persons in income and wealth do, of course, affect both life expectancy and quality of life. Statistical measures show economically disadvantaged groups have lower life expectancies and lower qualities of life on average. Nevertheless, neither of these differences of economic inequalities on total expected life years and their quality is likely to be as great with the QALY approach as with willingness to pay.

21 This notion that in death each person suffers the same loss, and moreover loses 'everything' by permanently ceasing to exist, is one important source of the view that when not all can be saved each should be given an equal chance of being selected to be saved, without differentiating for life years, quality of life years, value to others, and so forth. In what form, if any, this egalitarian intuition survives critical scrutiny is an important question, though not one that I can pursue here.

22 Lionel Needleman: 1976, 'Valuing other people's lives,' *The Manchester School of Economics and Social Studies* 44, no. 4, pp. 309–342; and M. W. Jones-Lee: 1976, *The Value of Life: An Economic Analysis* (University of Chicago Press, Chicago).

23 P. Menzel, *op. cit.,* pp. 40–42.

Chapter 10

Quality of life measures in health care and medical ethics

INTRODUCTION

There has been considerable philosophical work during the last two decades, especially in the United States but not limited to there, in a relatively new field called medical ethics. My aim in this essay is to explore what illumination that body of work might offer to our understanding of the quality of life. If one looks only to the medical ethics literature explicitly addressing the notion of the quality of life, there are few sustained analyses of it and of its role in various medical and health care contexts. Consequently, it is necessary to look more broadly to issues and areas of research that often do not explicitly address the quality of life, but that nevertheless have an important bearing on it. I believe there are two main areas of work in medical ethics that fit this criterion. The first is work on ethical frameworks for medical treatment decision-making in a clinical context, including accounts of informed consent and life-sustaining treatment decisions. The second is the development of valuational measures of outcomes of health care treatments and programs; these outcome measures are designed to guide health policy and so must be able to be applied to substantial numbers of people, including across or even between whole societies. The two main parts of this paper will address these two main bodies of work. Before doing so, however, several preliminary issues need to be briefly addressed.

I have mentioned that the literature that I will be summarizing and drawing on often does not explicitly address the concept of the 'quality of life', but instead uses other notions that are either closely related or roughly equivalent in the context. Sometimes a notion of 'health' is employed, particularly in its broader interpretations, as exemplified in the World Health Organization definition of 'health' as a state of complete physical, mental, and social well-being.[1] The notion of patient 'well-being', independent of its use within a definition of health, is also often employed for evaluation of outcomes in health care. Another conceptual framework commonly employed for evaluating health care outcomes is the assessment of the benefits and burdens of that care for the patient (and sometimes for others as well). Still another common conceptual framework often employed looks to the effects of health care on patients' interests, with a best interests standard particularly prominent for patients whose preferences cannot be determined. These and other conceptual schemes are not fully interchangeable in health care, much less in broader contexts. Nevertheless, they all have in common their use in evaluating health care outcomes for patients and their employment as at least part of a comprehensive account of a good life for persons. I shall freely draw here on each of these conceptual frameworks, and others, though indicating where differences between them become important.

The 'quality of life' can be given a number of more or less broad interpretations, depending on the scope of the evaluative factors concerning a person's life that it is taken to include. Medicine and health care often affect a person's life in only some limited areas or respects. Nevertheless, my concern will be with the broadest conception of, in Derek Parfit's words, 'what makes a life go best', and I shall try to show that medicine and health care may affect and illuminate more aspects of that question than might at first be thought.[2]

No concept is entirely apt or widely accepted in either philosophical or common usage for this broad role, but I shall use the concept of a 'good life' to refer to the quality of life of persons in its broadest interpretation.

It is common in much philosophical work on theories of the good for persons or of a good life to distinguish three broad kinds of theory. While this classification misses some distinctions important for my purposes here, it provides a natural starting point. These three alternative theories I will call the hedonist, preference satisfaction, and ideal theories of a good life.[3] Much of the philosophical work on these theories has been in the service of developing an account of 'utility', broadly construed for employment in consequentialist moral theories.[4] What is common to hedonist theories, as I will understand them here, is that they take the ultimate good for persons to be the undergoing of certain kinds of conscious experience. The particular kinds of conscious experience are variously characterized as pleasure, happiness, or the satisfaction or enjoyment that typically accompanies the successful pursuit of our desires. Particular states of the person that do not make reference to conscious experience, such as having diseased or healthy lungs, and particular activities, such as studying philosophy or playing tennis, are part of a good life on this view only to the extent that they produce the valuable conscious experience.

Preference satisfaction theories take a good life to consist in the satisfaction of people's desires or preferences. I here understand desires or preferences as taking states of affairs as their objects: for example, my desire to be in Boston on Tuesday is satisfied just when the state of affairs of my being in Boston on Tuesday obtains. This is to be distinguished from any feelings of satisfaction, understood as a conscious experience of mine, that I may experience if I am in Boston on Tuesday. The difference is clearest in cases in which my desire is satisfied, but I either do not or could not know that it is and so receive no satisfaction from getting what I desire: for example, my desire that my children should have long and fulfilling lives, a state of affairs that will only fully obtain after my death. For preference satisfaction theories of a good life to be at all plausible, they must allow for some correcting or 'laundering' of a person's actual preferences.[5] The most obvious example is the need to correct for misinformed preferences: for example, my desire to eat the sandwich before me not knowing that its ingredients are bad and will make me ill. Other corrections of preferences have also been supported by proponents of the preference satisfaction theory that are compatible with its underlying idea that ultimately what is good for persons is that they should get what they most want or prefer.

The third kind of theory holds that at least part of a good life consists neither of any conscious experience of a broadly hedonist sort nor of the satisfaction of the person's corrected preferences or desires, but instead consists of the realization of specific, explicitly normative ideals.[6] For example, many have held that one component of a good life consists in being a self-determined or autonomous agent, and that this is part of a good life for a person even if he or she is no happier as a result and has no desire to be autonomous. Ideal theories will differ both in the specific ideals the theories endorse and in the place they give to happiness and preference satisfaction in their full account of the good for persons. There is a strong tendency in much of the philosophical literature to seek a simple, comprehensive theory, such as the hedonist or preference satisfaction theories: proponents of ideal theories commonly acknowledge a plurality of component ideals that place constraints on and/or supplement the extent to which happiness and/or preference satisfaction serves a person's good. The account I will develop of quality of life judgments in health care strongly suggests that it is a mistake to let the attractions of a simple, unified theory of a good life force a choice between the hedonist and preference satisfaction theories. Instead, these quality of life judgments suggest the importance of giving independent place to the considerations singled out by each of the three main alternative theories, as ideal theories do, in any adequate overall account of the quality of life or of a good life for persons. The quality of life judgments made in medicine and health care also help some to fill in the content of a theory of a good life.

A major issue concerning ethical judgments generally, and judgments concerning a good life in particular, is the sense and extent to which such judgments are objective or subjective. A number of different senses have been given to the notions of 'objectivity' and 'subjectivity' in these contexts.[7] I will not attempt an extended analysis of this general theoretical issue here. Nevertheless, one sense in which what constitutes a good life for a particular person is believed to be subjective or objective mirrors the distinction between hedonist and preference satisfaction theories on the one hand, and ideal theories on the other. Hedonist and preference theories are both subjective in the sense that both hold that what is good for a particular person depends on what in fact makes that person happy or what that person in fact (with appropriate corrections) desires. (This is compatible, of course, with acknowledging that what will make a particular person happy or satisfy his or her preferences is an 'objective matter of fact', even if often an extremely difficult one to determine.)

Ideal theories are objective, or at least contain objective components, in the sense that they hold that a good life for a person is, at least in part, objectively determined by the correct or justified ideals of the good life, and does not in those respects depend either on what makes that person happy or on what that person's (even corrected) preferences happen to be. The question of whether accounts of a good life are objective or subjective is, then, an explicitly normative issue about what is the correct or most justified substantive theory of a good life. This sense of the objective–subjective dispute has been a central concern in the debates in medical ethics and health care about the quality of life. Interestingly, I believe that medicine and health care provide some of the most persuasive instances for both the objective *and* the subjective components of a good life, and so point the way towards a theory that incorporates hedonist, preference, and ideal components.

Haavi Morreim has distinguished a different sense in which quality of life judgments in medicine are either objective or subjective.[8] In her account, objective quality of life judgments are made on the basis of intersubjectively observable, material facts about a person (facts concerning his or her body, mind, functional capabilities, and environment), together with a socially shared evaluation of those facts, specifically of how those facts determine the person's quality of life. Subjective quality of life judgments also appeal to material facts about a person and his or her condition (though these may also include facts about the person's private psychological states), together with *that person's* value judgments about how those facts affect his or her quality of life. According to this account, the essential issue that determines whether a quality of life judgment is objective or subjective is whether the evaluative judgments concerning a particular individual's quality of life are and must be shared by some wider group or are, instead, only the individual's own. Since there are many possible wider social groups, one respect in which one could make sense of degrees of this kind of objectivity is in terms of the size, breadth, or nature of the wider social group; important variants include an individual's community or larger society, and a maximally wide group might be all humans or rational agents. It should be obvious that my and Morreim's senses of the objective–subjective distinction are independent: the individual whose quality of life is in question might hold any of the three substantive theories of the good for persons distinguished above, as might any wider social group.

A full conception of a good life for a person that does not reduce to a single property like happiness or preference satisfaction must assign a weight to the various components that contribute to that life's being good, though there may not be full comparability between different components and so in turn only partial comparability between different possible life courses for a person. Amartya Sen has suggested in several places the formal device of understanding these different components as independent vectors, each of which contributes to an overall assessment of the degree to which a person has a good life.[9] There are several benefits to an analysis of what constitutes a good life into a number of independent vectors. First, it allows us to accept part of what proponents of each of the three traditional theories of a good life have wanted to insist on, namely the theoretical independence of those components. The three components of happiness, preference satisfaction, and ideals of a good life can each be represented by their individual vectors, or subdivided further into distinct vectors within each component, having independent weight within an overall account of a good life. Second, the vector approach quite naturally yields the possibility of two senses of partial comparability of the quality of different lives. For a single individual, alternative possible lives may be only partially commensurable if one alternative life provides a greater value on one vector, but a lesser value on another vector, than another possible life. But for two different persons it is important that at least partial comparability between their lives may be possible, contrary to the dogma about the impossibility of interpersonal comparisons of utility, by comparing common vectors or by comparing different changes in common vectors making up a good life for each. Medicine and health care provide strong grounds for insisting on these independent vectors and, perhaps more important, also suggest a content and structure to the ideals along the lines proposed by Sen in his work on agency and capabilities, which drew on settings largely outside of health care.

We also need to distinguish between the relative importance of a particular feature or condition, say as represented by a specific vector, in its contribution to a person having a good life, compared with what I shall call its broader moral importance. A simple example will suffice. One condition that may plausibly contribute to a person's quality of life or good life is his or her physical mobility. It may be possible to specify roughly a normal level of physical mobility for persons of a similar age at a particular historical stage and in a particular society, and then to specify roughly levels of mobility say 25 per cent below and 25 per cent above the norm, such that the effect on a person's quality of life in moving from 25 per cent below the norm up to the norm is quantitatively roughly the same as moving from the norm to 25 per cent above it. While the degree or importance of the two changes in a person's quality of life or good life may be roughly the same, it can none the less be consistently held that these two comparable effects on the person's quality of life have different *moral* importance or priority. It might be held, for example, that on grounds of equality of opportunity bringing a person's mobility from 25 per cent below the norm up to the norm has greater moral priority than increasing his mobility from the norm to 25 per cent above it. The general point is that aspects of a person's quality of life may play a role not only in judgments about his quality of life or about how good a life he has, but also in other distinct moral and political judgments, or in the application of independent moral principles such as a principle of equal opportunity. This is, of course, a thoroughly familiar point in moral and political philosophy generally, and concerning consequentialist moralities in particular, against which it is often objected that they ignore the moral importance of whether the good is fairly or justly distributed. In the present context its importance is in reminding us to distinguish judgments concerning the improvement or reduction of people's quality of life from other independent moral evaluations of those same changes so as not to confuse needlessly the nature of quality of life judgments in health care.

ETHICAL FRAMEWORKS FOR HEALTH CARE TREATMENT DECISION-MAKING

The first broad area of work within medical ethics bearing on the concept of the quality of life concerns the aims of medicine and the account of medical treatment decision-making appropriate to those aims. It may be helpful to begin with a natural objection to thinking that these issues in medical ethics will illuminate any broad notion of the good life. On the contrary, as Leon Kass has argued, medicine's proper end is the much narrower one of health, or the healthy human being, and other goals such as happiness and gratifying patient desires are false goals for medicine.[10] Kass understands health to be a naturalistically defined property of individual biological organisms, organisms which must be understood as organic wholes, and whose parts

> have specific functions that define their nature as parts: the bone marrow for making red blood cells; the lungs for exchange of oxygen and carbon dioxide; the heart for pumping the blood. Even at a biochemical level, every molecule can be characterized in terms of its function. The parts, both macroscopic and microscopic, contribute to the maintenance and functioning of the other parts, and make possible the maintenance and functioning of the whole.[11]

What constitutes well-functioning varies with the particular biological species in question, but Kass is at pains to argue that 'health is a natural standard or norm – not a moral norm, not a "value" as opposed to a "fact", not an obligation – a state of being that reveals itself in activity as a standard of bodily excellence or fitness'.[12]

Kass's work constitutes one of the more ambitious attempts to justify two common-sense beliefs about the 'objectivity' of medicine: that the aim of medicine is and should be the patient's health, and that health is a biologically determined, objective matter of fact. If so, then physicians, with their impressive body of scientific knowledge concerning human biological functioning and the impact of therapeutic interventions on diseases and their natural courses, would seem to be the proper judges of whether we are healthy and, if we are not, what therapeutic interventions will be likely to make us more so. This hardly begins to do justice to the subtlety of Kass's view – though it is a view that I believe to be fundamentally mistaken – but it does bring out why one might think medicine, properly aimed only at human health defined in terms of biological functioning, has little to teach us regarding broader social issues about the quality of life. I believe it is fair to say that the main body of work in medical ethics within the last two decades has rejected Kass's view that the sole proper aim of medicine is health, defined in naturalistic, biological terms, and the ethical framework for medical treatment decision-making that it would seem to imply. We need to see how the alternative, broader view of the aims of medicine that should guide medical treatment decision-making bears on an understanding of the quality of life.

It has become a commonplace, at least in the developed countries, that medicine has achieved the capacity commonly to offer to patients suffering from particular diseases a number of alternative treatments, and to extend patients' lives in circumstances in which the benefit to the patient of doing so is increasingly problematic. In the United States this has led to patients pursuing various means of gaining control over decisions about their treatment. In the case of competent patients, a broad consensus has developed that such patients have the right to decide about their care in a process of shared decision-making with their physicians and to reject any proffered treatment. In the case of incompetent patients, an analogous consensus has been developing that an incompetent patient's surrogate, seeking to decide as the patient would have decided in the circumstances if competent, is likewise entitled to decide about the patient's care with the patient's physician and to reject any care the patient would not have wanted – though the consensus concerning incompetent patients is less broad and more ringed with qualifications. Each consensus is reflected in a large medical ethics literature, a growing body of legal decisions, legal mechanisms such as Living Wills and Durable Powers of Attorney for Health Care, whose purpose is to ensure patients' control over their care, pronouncements and studies of authoritative bodies and commissions, policies of health care institutions, and the practice of health care professionals.[13]

`The common view now is that health care decision-making should be a process of shared decision-making between patient (or the patient's surrogate in the case of an incompetent patient) and physician.[14] Each is seen as indispensable to sound decision-making. The physician brings his or her training, knowledge, and expertise to bear for the diagnosis of the patient's condition, the estimation of the patient's prognosis with different alternative treatments, including the alternative of no treatment, and a recommendation regarding treatment. The patient brings the knowledge of his or her aims, ends, and values that are likely to be affected by different courses of treatment, and this enables a comparative evaluation of different possible outcomes to be made. As alternative treatments have multiplied and become possible in circumstances promising increasingly marginal or questionable benefits, both physicians and patients are called upon to make increasingly difficult judgments about the effects of treatment on patients' quality of life. It is worth noting that proponents of shared decision-making need not reject the functional account of health as a biological norm defended by Kass and others. What they can reject is the claim that the only proper goal of medicine is health. Instead, medicine's goal should be to provide treatment that best enables patients to pursue successfully their overall aims and ends, or life plans. It is the relative value of health, and of different aspects of health, as compared with other ends, that varies for different persons and circumstances.

Most patients' decisions about life-sustaining treatment will be based on their judgment of the benefits and burdens of the proposed treatment and the life it sustains, though in some instances patients may give significant weight to other factors such as religious obligations, the emotional burdens and financial costs for their families, and so forth. Except for patients who hold a form of vitalism according to which human life should or must be sustained at all costs and whatever its quality, these decisions by competent patients must inevitably involve an assessment of their expected quality of life if life-sustaining treatment is employed, though, as I shall note shortly, of only a very restricted sort.

Some have rejected the acceptability of quality of life judgments in the case of incompetent patients unable to decide for themselves, for whom others must therefore make treatment decisions.[15] One version of the objection is that no one should decide for another whether that other's quality of life is such that it is not worth continuing it. More specifically, the objection is that it is unacceptable to judge that the quality of another person's life is so poor that it is not worth the cost and effort to others to sustain that person's life. This objection, however, is not to making quality of life judgments in this context generally, but only to concluding that a person's life is not worth sustaining because its poor quality makes it not of value, but instead a burden, to *others*. The sound point that this objection confusingly makes is that quality of life judgments concerning a particular person should address how the conditions of a person's life affect its quality or value to *that person*, and not its value to others. Moreover, persons might judge their quality of life to be low and nevertheless value their lives as precious. In economic and policy analysis one version of the so-called human capital method of valuing human life, which values a person's life at a given point in time by his or her expected future earnings minus personal consumption, in effect values a person's life in terms of its economic value to others.[16] But there is no reason to reject the soundness of any evaluation by one person of another's quality of life simply because some might draw a further unjustified conclusion that if its quality is sufficiently low to make it on balance a burden to others, it ought not to be sustained.

The quality of life judgment appropriate to life-sustaining treatment decisions, whether made by a competent patient or an incompetent patient's surrogate, should thus assess how the conditions of the patient's life affect the value of that life to that patient. Nevertheless, even properly focused in this way, the role of quality of life judgments in decisions about whether to withhold or withdraw life-sustaining treatment is extremely limited. This quality of life judgment focuses only on which side of a *single threshold* a person's quality of life lies. The threshold question is: 'Is the quality of the patient's life so poor that for that person continued life is worse than no further life at all?' Or, in the language of benefits and burdens commonly employed in this context: 'Is the patient's quality of life so poor that the use of life-sustaining treatment is unduly burdensome, that is, such that the burdens to the patient of the treatment and/or the life that it sustains are sufficiently great and the benefits to the patient of the life that is sustained sufficiently limited, to make continued life on balance no longer a benefit or good to the patient?'[17] The only discrimination in quality of life required here is whether the quality of the life is on balance sufficiently poor to make it worse than non-existence to the person whose life it is.

Some have objected that this judgment is incoherent since, though it is possible to compare the quality of two lives, or of a single person's life under different conditions, it is not possible to make the quality of life comparison needed here because one of the alternatives to be compared is non-existence. If a person no longer exists, there is no life that could possibly have any quality so as to enter into a comparison with the quality of the life sustained by treatment. This objection does correctly point out that the judgment in question cannot involve a comparison of the quality of two alternative periods of life, though it could compare two possible lives, one that ends at that time and another that continues longer. However, it does not follow that there is no sense to the question of whether the best life possible for a person with some form of life-sustaining treatment is of sufficiently poor quality, or sufficiently burdensome, to be worse for that person than no further life at all. Perhaps the most plausible example is the case of a patient suffering from an advanced stage of invariably fatal cancer, who is virtually certain to die within a matter of days whatever is done, and whose life will be filled in those remaining days with great and unrelievable pain and suffering. (With the appropriate use of presently available measures of pain relief, it is in fact only very rarely the case that great pain and suffering in such cases cannot be substantially relieved.) The burden of those remaining days may then be found by the patient to be virtually unbearable, while the life sustained provides nothing of value or benefit to the patient. This judgment addresses the quality of the life sustained and appears to be a sensible judgment. It is just the judgment that patients or their surrogates commonly understand themselves to be making when they decide whether to employ or continue life-sustaining treatment.

Alternatively, the objection to someone ever making such a judgment for another may not be based on any putative incoherence of such judgments, but may express instead a concern about the difficulty of ever reliably deciding how *another* would in such circumstances decide, due perhaps to the diversity and unpredictability of people's actual decisions for themselves. Moreover, if the difficulty of reliably making such judgments for others is in fact this great, then we might well have a related practical concern that the interests of others, which may be in conflict with those of the patient, may consciously or unconsciously infect judgments about what is best for the patient.

Despite these difficulties, there have been attempts to formulate some general substantive standards to determine when an incompetent patient's quality of life is so poor that withholding or withdrawing further life-sustaining treatment is justified. Nicholas Rango, for example, has proposed standards for nursing home patients with dementia.[18] He emphasizes the importance of being clear about the purposes for which care is provided and distinguishes three forms of care: (1) palliative care aimed at relieving physical pain and psychological distress; (2) rehabilitative care aimed at identifying and treating 'excess disabilities, the gap between actual level of physical, psychological or social functioning and potential functioning capacity';[19] and (3) medical care aimed at reducing the risk of mortality or morbidity. He emphasizes the importance of therapeutic caution because a seriously demented patient will not be able to understand the purposes of painful or invasive interventions, and so presumably cannot choose to undergo and bear burdensome treatment for the sake of promised benefits.[20] Rango proposes two conditions, either of which is sufficient to justify forgoing further treatment of a chronic medical condition or a superimposed acute illness: (*a*) when the patient is burdened by great suffering despite palliative and rehabilitative efforts; (*b*) when the dementia progresses 'to a stuporous state of consciousness in which the person lives with a negligible awareness of self, other, and the world'.[21] Even within the relatively narrow focus of life-sustaining treatment decisions for demented patients, Rango's proposal can be seen to include three different kinds of components of quality of life assessments. The first, covered by treatment aim (1) and patient conditions (*a*) and (*b*), concerns the quality of the patient's conscious experience. The second, covered especially by treatment aim (2) and patient condition (*b*), concerns the patient's broad functional capacities. The third, covered especially by patient condition (*b*) and by the patient's ability to understand the purpose of treatment and in turn to choose to undergo it, concerns the centrality to quality of life of the capacity to exercise choice in forming and pursuing an integrated and coherent life plan. I shall argue below that each of these three kinds of condition is an essential component of an adequate account of the quality of life.

At the other end of life, the debate in the United States about treatment for critically ill new-born infants has also focused on the role of quality of life considerations in determining when life-sustaining treatment is a benefit for the infant. One influential attempt to bring quality of life considerations into these decisions is the proposal of the moral theologian, Richard McCormick, that a new-born infant's life is a value that must be preserved only if the infant has the potential for a 'meaningful life'.[22] A meaningful life is one that contains some potential for human relationships: anencephalic new-borns, for example, wholly lack this potential, while those with Down's syndrome or spina bifida (to cite two of the most discussed kinds of case) normally do not. Nancy Rhoden has developed a more detailed proposal along similar lines regarding life-sustaining treatment for new-born infants:

> aggressive treatment is not mandatory if an infant: (1) is in the process of dying; (2) will never be conscious; (3) will suffer unremitting pain; (4) can only live with major, highly restrictive technology which is not intended to be temporary (e.g., artificial ventilation); (5) cannot live past infancy (i.e., a few years); or (6) lacks potential for human interaction as a result of profound retardation.[23]

Rhoden's proposal is typical of those accepting the use of quality of life considerations in decisions about life-sustaining treatment for new-born infants in its focus on the infant's at least minimal capacity for positive conscious experience (conditions 2 and 3): the infant's capacities for physical, mental, and social functioning (conditions 2, 3, 4, and 6), and the infant's capacities to live far enough into childhood to begin a life that can be viewed and experienced 'from the inside' by the child as a life lived in a biographical, not biological, sense (conditions 1 and 4; I will say more about this special feature of infant 'life years' below in discussing the relevance of mortality data to good lives). However, the very limited, single-threshold character of the quality of life assessments required in decisions as to whether to forgo or to employ life-sustaining treatment, whether for adults or new-born infants, takes us only a little way in understanding quality of life assessments in health care treatment decision-making.

It is necessary, consequently, to broaden the focus from life-sustaining treatment decisions to medical treatment decisions generally. Here, as noted earlier, there is a widespread consensus that competent patients are entitled, in a process of shared decision-making with their physicians, to decide about their treatment and to refuse any proffered or recommended treatment. In the United States, the doctrine of informed consent, both in medical ethics and in the law, requires that treatment should not be given to a competent patient without that patient's informed and voluntary consent.[24] What does this doctrine, which lodges decision-making authority with the patient, imply about the nature of judgments concerning the patient's quality of life? An argument that it presupposes the normatively subjective, preference satisfaction account of a good life might, in rough outline, go as follows.

Each of the requirements of the informed consent doctrine can be understood as designed to provide reasonable assurance that the patient has chosen the treatment alternative most in accord with his or her own settled preferences and values. If the patient's decision is not informed – specifically, if the patient is not provided in understandable form with information regarding his or her diagnosis, the prognosis when different treatment alternatives (including the alternative of no treatment) are pursued, including the expected risks and benefits, with their attendant probabilities, of treatment alternatives – then the patient will lack the information needed to select the alternative most in accord with his or her settled preferences. If the patient's decision is not voluntary, but is instead forced, coerced, or manipulated by another, then it is likely not to be in accordance with the patient's settled preferences, but instead will forward another's interests, or another's view of what is best for the patient. If the patient is not competent to make the choice in question, then he or she will lack the ability to use the information provided to deliberate about the alternatives and to select the one most in accord with his or her settled preferences. When all three requirements are satisfied – the decision is informed and voluntary, and the patient competent – others can be reasonably assured that the patient's choice fits the patient's own conception of a good life, as reflected in his or her settled preferences.[25] Viewed in this way, the informed consent doctrine may appear to be grounded in a preference satisfaction account of the good or the quality of life, and so not to require any more complex vector account of the sort suggested earlier. Even in this very crude form, however, this argument can be seen to be unsound if one asks what values the informed consent doctrine and the account of shared decision-making in medicine are usually thought to promote, and what values support their acceptance.

The most natural and obvious first answer has already implicitly been given: the informed consent doctrine in health care treatment decision-making is designed, when its three requirements are satisfied, to serve and promote the patient's well-being, as defined by the patient's settled, uncoerced, and informed preferences. If this were the only value at stake, or at least clearly the dominant value, then it would be plausible to argue that the informed consent doctrine rests, at least implicitly, on a preference satisfaction account of the good life. However, it is not the only value at stake. Usually regarded as at least of roughly commensurate importance is respecting the patient's self-determination or autonomy.[26] The interest in self-determination I understand to be the interest of persons, broadly stated, in forming, revising over time, and pursuing in their choices and actions their own conception of a good life; more narrowly stated for my specific purposes here, it is people's interest in making significant decisions affecting their lives, such as decisions about their medical care, for themselves and according to their own values. Sometimes this is formulated as the right to self-determination.

Whether interest or right, however, the greater the moral weight accorded to individual self-determination as one of the values underlying and supporting the informed consent doctrine, the weaker the basis for inferring that the doctrine presupposes the normatively subjective, preference satisfaction account of the good life. This is because the greater the moral weight accorded to individual self-determination, the more self-determination can explain the requirement of informed consent, even assuming the patient chooses in a manner sharply contrary to his or her own well-being. Moreover, there is substantial reason to suppose that the doctrine does in fact largely rest on respect for patients' self-determination. In the celebrated 1914 legal case of *Schloendorff* v. *Society of New York Hospital,* usually cited as the first important enunciation of the legal requirement of consent for medical care, Justice Cardozo held that: 'Every human being of adult years and sound mind has a right to determine what shall be done to his own body; and a surgeon who performs an operation without his patient's consent commits an assault, for which he is liable in damages.'[27] I shall make no attempt here to trace the development of the legal doctrine of informed consent since *Schloendorff,* but it is probably fair to say that no other subsequent case has been as influential or as often cited in that development. And the later cases, in one form or another, repeatedly appeal to a right to self-determination to support that developing doctrine.[28] Nearly half a century later, for example, in the important 1960 case of *Natanson* v. *Kline* the Kansas Supreme Court made an equally ringing appeal to self-determination:

> Anglo-American law starts with the premise of thorough-going self-determination. It follows that each man is considered to be master of his own body, and he may, if he be of sound mind, expressly prohibit the performance of life-saving surgery, or other medical treatment. A doctor might well believe that an operation or form of treatment is desirable or necessary but the law does not permit him to substitute his own judgment for that of the patient by any form of artifice or deception.[29]

The philosophical tradition regarding the problem of paternalism is equally bound up in a commitment to the importance of individual self-determination or autonomy. Here, the *locus classicus* is John Stuart Mill's renowned assertion of the 'one very simple principle' that

> the sole end for which mankind are warranted, individually or collectively, in interfering with the liberty of action of any of their number is self-protection. That the only purpose for which power can be rightfully exercised over any member of

a civilized community, against his will, is to prevent harm to others. His own good, either physical or moral, is not a sufficient warrant. He cannot rightfully be compelled to do or forbear because it will be better for him to do so, because it will make him happier, because, in the opinions of others, to do so would be wise or even right.[30]

The voluminous subsequent philosophical literature on paternalism certainly suggests that this principle is not as simple as Mill supposed, but it also makes clear that, even for one like Mill who in other contexts was an avowed utilitarian, the case for non-interference with individual self-determination and liberty of action does not rest on any claim that doing so cannot be for a person's 'good', understood in a normatively subjective interpretation. Quite to the contrary, Mill and the many who have followed him have been at pains to insist that such interference is not justified even when it would truly be for the good of the one interfered with.[31] The result is that it is not possible to draw any firm conclusion from the doctrine of informed consent in medicine that patients' well-being or quality of life is understood according to a normatively subjective interpretation. Individual self-determination can serve as the foundation for the informed consent doctrine and can make that doctrine compatible with any of the three main alternative accounts of the good or quality of life that I have distinguished.

What is the relation between these two values of patient self-determination and well-being, commonly taken as underlying the informed consent doctrine, and the broad concept of a good life? The conventional view, I believe, is that the patient's well-being is roughly equivalent to the patient's good and that individual self-determination is a value independent of the patient's well-being or good. Respecting the patient's right to self-determination, then, at least sometimes justifies respecting treatment choices that are contrary to the patient's well-being or good.[32] Respecting self-determination is commonly held to be what is required by recognizing the individual as a person, capable of forming a conception of the good life for him or herself. If personal self-determination is a fundamental value – fundamental in that it is what is involved in respecting persons – however, then I suggest that our broadest conception of a good life should be capable of encompassing it rather than setting it off as separate from and in potential conflict with a person's well-being or good, as in the conventional account of informed consent. What we need is a distinction between a good life for a person in the broadest sense and a person's personal well-being, such that only personal well-being is independent of and potentially in conflict with individual self-determination.

We should think of being self-determined as central to – a central part of – having a good life in the broadest sense. It is in the exercise of self-determination that we maintain some control over, and take responsibility for, our lives and for what we will become. This is not to deny, of course, that there are always substantial limits and constraints within which we must exercise this judgment and choice. But it is to say that showing respect for people through respecting their self-determination acknowledges the fundamental place of self-determination in a good life. I take this to be essentially what Rawls intends by his claim that people have a highest-order interest in autonomy and what Sen means by his notion of agency freedom.[33] We do not want this broad conception of a good life, however, to prevent our making sense of persons freely and knowingly choosing to sacrifice their own personal well-being for the sake of other persons. To cite an extreme case, a parent might knowingly and freely choose not to pursue expensive life-sustaining treatment such as a heart transplant in order to preserve financial resources for his children's education. Given his love and sense of responsibility for his children, he would judge his life to be worse if he had the transplant at the expense of his children's education, though certainly his health and his personal well-being would be improved. We might say here that he values his personal well-being in these circumstances less than the well-being of his children. In its requirement that such a choice be respected, the informed consent doctrine implicitly accepts that the best life for a person is a life of self-determination or choice, even if the exercise of that self-determination or choice results in a lessened state of personal well-being. Precisely how to make this distinction between the good life, as opposed to the personal well-being, of an individual raises difficulties that I cannot pursue here. The rough idea is that personal well-being makes essential reference to the states of consciousness, activities, and capacities for functioning of the person in question (I will pursue this further in Section 3 of this chapter), and it is these that are worsened when the parent pursues his conception of a good life in sacrificing his personal well-being for that of his child.[34]

Medical treatment decisions must often be made for patients who are not themselves competent to make them. There has been considerable discussion both in medical ethics and in the law regarding appropriate ethical standards for such decisions.[35] Since quality of life considerations are virtually always relevant to these decisions, the ethical frameworks developed for these decisions must employ, either explicitly or implicitly, a conception of the quality of life of patients. There is considerable consensus that if the patient in question, while still competent, formulated and left an explicit advance directive clearly and unambiguously specifying his or her wishes regarding treatment in the circumstances now obtaining, then those wishes should be followed, at least within very broad limits, by those treating the patient. At the present time in the United States, most states have adopted legislation giving the force of law to one or another form of so-called Living Will, which allows people to give binding instructions about their treatment should they become incompetent and unable to decide for themselves. Several other states have more recently enacted legislation permitting people to draw up a Durable Power of Attorney for Health Care, which combines the giving of instructions about the person's wishes regarding treatment with the designation of who is to act as surrogate decision-maker, and so to interpret those instructions, should one become incompetent to make the decisions oneself.

In the usual case in which an incompetent patient has left no formal advance directive, two principles for the guidance of those who must decide about treatment for the patient have been supported – the *substituted judgment* principle and the *best interests* principle. The substituted judgment principle requires the surrogate to decide as the patient would have decided if competent and in the circumstances that currently obtain. The best interests principle requires the surrogate to make the treatment decision that best serves the patient's interests. This has the appearance of a dispute between what I earlier called normatively subjective and normatively objective accounts of a good life, since the only point of the best interests principle as an alternative to substituted judgment might seem to be that it employs a normatively objective standard of the person's good that does not depend on his or her particular subjective preferences and values. However, this appearance is misleading. These two principles of surrogate decision-making are properly understood, in my view, not as competing alternative principles to be used for the same cases, but instead as an ordered pair of principles to cover all cases of surrogate decision-making for incompetent patients in which an advance directive does not exist, with each of the two principles to apply in a different subset of these cases. (This is not to say that these two principles are always in fact understood in medical ethics, the law, or health care practice as applying to distinct groups of cases; the treatment of these two principles is rife with confusion.)

The two groups of cases are differentiated with regard to the information available or obtainable concerning the patient's general preferences and values that has some bearing on the treatment choice at hand. The two principles are an ordered pair in the sense that when sufficient information is available about the relevant preferences and values of the patient to permit a reasonably well-grounded application of the substituted judgment principle, then surrogate decision-makers for the patient are to use that information and that principle to infer what the patient's decision would have been in the circumstances if he or she were competent.[36] In the absence of such information, and only then, surrogate decision-makers for the patient are to select the alternative that is in the best interests of the patient, which is usually interpreted to mean the alternative that most reasonable and informed persons would select in the circumstances. Thus, these two principles are not competing principles for application in the same cases, but alternative principles to be applied in different cases.

Nevertheless, it might seem that the best interests standard remains a normatively objective account when it is employed. However, this need not be so. If the best interests standard is understood as appealing to what most informed and reasonable persons would choose in the circumstances, it employs the normatively subjective preference standard. And it applies the choice of most persons to the patient in question because in the absence of any information to establish that the patient's relevant preferences and values are different than most people's, the most reasonable presumption is that the patient is like most others in the relevant respects and would choose like those others. Thus, the best interests standard, like the two other standards of choice for incompetent patients – the advance directive and substituted judgment standards – can be understood as requiring the selection of the alternative the patient would most likely have selected, with the variations in the standards suited to the different levels of information about the patient that is available. Just as with the informed consent doctrine that applies to competent patients, so these three principles – advance directives, substituted judgment, and best interests – guiding surrogate choice for incompetent patients can all be understood as supported and justified by the values of patient well-being *and* self-determination. Thus, each of these three principles implicitly employs an account of a good life that is a life of choice and self-determination concerning one's aims, values, and life plan.

Before leaving the ethical frameworks that have been developed in the medical ethics, legal, and medical literatures for treatment decision-making for competent and incompetent patients, I want to make explicit an indeterminacy in these frameworks concerning the nature of the ethical theory they presuppose. I have noted that it is common to base these ethical frameworks on two central values in a good life: patient well-being and self-determination. What is commonly left unclear, however, is the foundational status of the ethical value of individual self-determination in the underlying ethical theory supporting these decision-making frameworks. Self-determination might be held to have only derivative or instrumental value within a broadly consequentialist moral theory. Specifically, it might be held to be instrumentally valuable for the fundamental value of happiness or preference satisfaction within normatively subjective theories of a good life of a hedonist or preference satisfaction sort. If, as seems true at least for many social conditions and historical periods, most people have a relatively strong desire to make significant decisions about their lives for themselves, then it is at least a plausible presumption that their doing so will generally promote their happiness or the satisfaction of their desires. If self-determination is only valuable at bottom in so far as it leads to happiness or preference satisfaction, then it will not be part of an ideal of the person that is objective in the sense of its value not being entirely dependent on happiness or preference satisfaction. Since on most plausible theories of the good, happiness and desire satisfaction are a significant *part* of a good life, and since self-determination does commonly make a significant contribution to people's happiness or desire satisfaction, self-determination will commonly have significant instrumental value on any plausible theory of a good life.[37]

As a result, to single out self-determination as one of the two principal values underlying the informed consent doctrine in medical ethics as it applies both to competent and incompetent patients is not to make clear at a foundational level of ethical theory whether self-determination is held to have only instrumental value, or also significant non-instrumental value as an important component of an objective ideal of the person. The vast majority of ethical discussions of informed consent and health care treatment decision-making simply do not either explicitly address this foundational question of ethical theory or even implicitly presuppose a particular position on it. From a practical perspective, this foundational indeterminacy has the value of allowing proponents of incompatible ethical theories, for example consequentialists and rights-based theorists, to agree on the fundamental importance of self-determination and choice in a good life.

There is one final difficulty to be noted in attempting to infer the account of a good life at the level of basic ethical theory from the ethical frameworks for treatment decision-making advocated for and employed in medical practice. There is a general difficulty in inferring the underlying values or ethical principles that support social practices. A social practice like that of informed consent and shared decision-making in medicine must guide over time a very great number of treatment decisions carried out in a wide variety of circumstances by many and diverse patients, family members, and health care professionals. A well-structured practice must take account of and appropriately minimize the potential in all involved parties for well-intentioned misuse or ill-intentioned abuse of their roles. Institutional constraints may thus be justified, though on some particular occasions they will produce undesirable results, because in the long run their overall results are better than those of any feasible alternatives. For example, even if we appeal only to the value of patient well-being and leave aside any independent value of self-determination, a strong right of patients to refuse any treatment might be justified if most of the time people are themselves the best judges of what health care treatment will best promote their happiness or satisfy their enduring preferences and values. Alternatively, that same strong right to refuse treatment might be justified within a normatively objective ideal theory of a good life for persons because, though individuals can be mistaken about their own good when they pursue their happiness or seek to satisfy their desires, no reliable alternative social and legal practice is feasible that will produce better results in the long run, even judged by that ideal theory of a good life. The imperfections and limitations of people and institutions may lead supporters of quite different accounts of a good life to support roughly the same institutions in practice. In a more general form, this is a thoroughly familiar point in moral philosophy where defenders of fundamentally different moral theories, such as consequentialists and rights theorists, may converge on the institutions justified by their quite different theories. Without explicitly uncovering the justificatory rationales for specific social institutions accepted by particular persons, we cannot confidently infer the ethical principles or judgments, and specifically the conception of a good life, that they presuppose.

HEALTH POLICY MEASURES OF THE QUALITY OF LIFE

I want now to shift attention from the account of the quality of life presupposed by ethical frameworks for medical treatment decision-making to more explicit measures used to assess health levels and the quality of life as it is affected by health and disease within larger population groups. Early measurement attempts focused on morbidity and mortality rates in different populations and societies. These yield only extremely crude comparisons, since they often employ only such statistics as life expectancy, infant mortality, and reported rates of specific diseases in a population. Nevertheless, they will show gross differences between countries, especially between economically developed and underdeveloped countries, and between different historical periods, in both length and quality of life as it is affected by disease. Major changes in these measures during this century, as is well known, have been due principally to public health measures such as improved water supplies, sewage treatment, and other sanitation programs and to the effects of economic development in improving nutrition, housing, and education; improvements in the quality of and access to medical care have been less important. In recent decades, health policy researchers have developed a variety of measures that go substantially beyond crude morbidity and mortality measures. Before shifting our attention to them, however, it is worth underlining the importance of mortality measures to the broad concept of a good life.

When quality and quantity of life are distinguished, both are relevant to the degree to which a person has a good life. People whose lives are of high quality, by whatever measure of quality, but whose lives are cut short well before reaching the normal life span in their society, have had lives that have gone substantially less well, because of their premature death, than reasonably might have been expected. People typically develop, at least by adolescence, more or less articulated and detailed plans for their lives; commonly, the further into the future those plans stretch, the less detailed, more general, and more open-ended they are. Our life plans undergo continuous revision, both minor and substantial, over the course of our lives, but at any point in time within a life people's plans for their lives will be based in part on assumptions about what they can reasonably expect in the way of a normal life span.[38] When their lives are cut short prematurely by illness and disease they lose not just the experiences, happiness, and satisfactions that they would otherwise have had in those lost years, but they often lose as well the opportunity to complete long-term projects and to achieve and live out the full shape, coherence, and conclusion that they had planned for their lives. It is this rounding out and completion of a life plan and a life that help enable many elderly people, when near death, to feel that they have lived a full and complete life and so to accept their approaching death with equanimity and dignity. The loss from premature death is thus not simply the loss of a unit of a good thing, so many desired and expected-to-be-happy life-years, but the cutting short of the as yet incompletely realized life plan that gave meaning and coherence to the person's life.

The importance of life plans for a good life suggests at least two other ways in which different mortality rates within societies affect the opportunities of their members to attain good lives. Citizens of economically underdeveloped countries typically have shorter life expectancies than do citizens of the developed countries. Thus, even those who reach a normal life span will have less time to develop and enjoy a richly complex and satisfying life than will those who reach a normal, but significantly longer, life span in the developed countries. Mortality data indicate that citizens of underdeveloped countries typically have less good lives as a result of inadequate health care *both* because of their shorter life expectancies *and* because of their increased risk of not living to even the normal life span in their own society.

One final relation between mortality data and the importance of life plans in a good life concerns infant and extremely early childhood mortality. It is common to view such early mortality as particularly tragic, both because of the greater amount of expected life-years lost to the individual and because the life was cut short just as it was getting started. The loss is often deep for the parents, in part because of the hopes and plans that *they* had for the infant's future. But the death of an infant or extremely young child before he or she has developed the capacity to form desires, hopes, and plans for the future cuts that life short – 'life' understood as a connected plan or unfolding biography with a beginning, middle, and end – before it has begun; the infant is alive, but does not yet have a life in this biographical sense.[39] From the perspective of this biographical sense of having a life as lived from the 'inside', premature death in later childhood, adolescence, or early adulthood commonly makes a life that has got started go badly, whereas infant death does not make a life go badly, but instead prevents it from getting started.

There is a voluminous medical and health policy literature focused on the evaluation of people's quality of life as it is affected by various disease states and/or treatments to ameliorate or cure those diseases.[40] The dominant conception of the appropriate aims of medicine focuses on medicine as an intervention aimed at preventing, ameliorating, or curing disease and its associated effects of suffering and disability, and thereby restoring, or preventing the loss of, normal function or of life. Whether the norm be that of the particular individual, or that typical in the particular society or species, the aim of raising people's function to *above* the norm is not commonly accepted as an aim of medicine of equal importance to restoring function *up to* the norm. Problematic though the distinction may be, quality of life measures in medicine and health care consequently tend to focus on individuals' or patients' *dysfunction* and its relation to some such norm. At a deep level, medicine views bodily parts and organs, individual human bodies, and people from a functional perspective. Both health policy analysts and other social scientists have done considerable work constructing and employing measures of health and quality of life for use with large and relatively diverse populations. Sometimes these measures explicitly address only part of an overall evaluation of people's quality of life, while in other instances they address something like overall quality of life as it is affected by disease. A closely related body of work focuses somewhat more narrowly on an evaluation of the effect on quality of life of specific modes of treatment for specific disease states. This research is more clinically oriented, though the breadth of impact on quality of life researchers seek to measure does vary to some extent, depending often on the usual breadth of impact on the person of the disease being treated. Generally speaking, the population-wide measures tend to be less sensitive to individual differences as regards both the manner and the degree to which a particular factor affects people's quality of life. It will be helpful to have before us a few representative examples of the evaluative frameworks employed.[41]

The Sickness Impact Profile (SIP) was developed by Marilyn Bergner and colleagues to measure the impact of a wide variety of forms of ill health on the quality of people's lives.[42] Table 10.1 enumerates the items measured.

A second example of an evaluative framework is the Quality of Life Index (QLI) developed by Walter O. Spitzer and colleagues to measure the quality of life of cancer patients (see Table 10.2).[43]

A third prominent measure developed by Milton Chen, S. Fanshel and others is the Health Status Index (HSI), which measures levels of function along certain dimensions (see Table 10.3).[44]

It would be a mistake, of course, to attempt to infer precise and comprehensive philosophical theories of the quality of life or of a good life from measures such as these. The people who develop them are commonly social scientists and health care researchers who are often not philosophically sophisticated or concerned with the issues that divide competing philosophical accounts of a good life. The practical and theoretical difficulties in constructing valid measures that are feasible for large and varied populations require compromises with and simplifications of – or simply passing over – issues of philosophical importance. Nevertheless, several features of these measures are significant in showing the complexity of the quality of life measures employed in health care and, I believe, of any adequate account of the quality of life or of a good life.

First, the principal emphasis in each of the three measures of quality of life is on function, and functions of the 'whole person' as opposed to body parts and organ systems. In each case the functions are broadly characterized so as to be relevant not simply to a relatively limited and narrow class of life plans, but to virtually any life plan common in modern societies. Following the lead of Rawls's notion of 'primary goods', I shall call these 'primary functions'.[45] In the SIP, the categories of sleep and rest, and eating are necessary for biological function. The categories of work, home management, and recreation and pastimes are central activities common in virtually all lives, though the relative importance they

Table 10.1. *Sickness Impact Profile Categories and Selected Items*

Dimension	Category	Items describing behaviour related to:	Selected items
Independent categories	SR	Sleep and rest	I sit during much of the day
			I sleep or nap during the day
	E	Eating	I am eating no food at all, nutrition is taken through tubes or intravenous fluids
			I am eating special or different food
	W	Work	I am not working at all
			I often act irritable towards my work associates
	HM	Home management	I am not doing any of the maintenance or repair work around the house that I usually do
			I am not doing heavy work around the house
	RP	Recreation and pastimes	I am going out for entertainment less
			I am not doing any of my usual physical recreation or activities
I. Physical	A	Ambulation	I walk shorter distances or stop to rest often
			I do not walk at all
	M	Mobility	I stay within one room
			I stay away from home only for brief periods of time

Table 10.1. (continued)

Dimension	Category	Items describing behaviour related to:	Selected items
	BCM	Body care and movement	I do not bathe myself at all, but am bathed by someone else I am very clumsy in my body movements
II. Psychosocial	SI	Social interaction	I am doing fewer social activities with groups of people I isolate myself as much as I can from the rest of the family
	AB	Alertness behaviour	I have difficulty reasoning and solving problems, e.g. making plans, making decisions, learning new things I sometimes behave as if I were confused or disoriented in place or time, e.g. where I am, who is around, directions, what day it is
	EB	Emotional behaviour	I laugh or cry suddenly I act irritable and impatient with myself, e.g. talk badly about myself, swear at myself, blame myself for things that happen
	C	Communication	I am having trouble writing or typing I do not speak clearly when I am under stress

Table 10.2. *Quality of Life Index: Format of the Final Version Adopted*

Study No _____

Age _____
See M_1 F_2 (Ring appropriate letter)
Primary Problem or Diagnosis _____

Secondary Problem or Diagnosis, or complication (if appropriate) _____
Scorer's Specialty _____

☐☐
☐

☐☐☐
☐☐☐

Scoring Form

☐

Score each reading 2, 1 or 0 according to your most recent assessment of the patient.

ACTIVITY   *During the last week, the patient*

- has been working or studying full time, or nearly so, in usual occupation; or managing own household, or participating in unpaid or voluntary activities, whether retired or not 2
- has been working or studying in usual occupation or managing own household or participating in unpaid or voluntary activities, but requiring major assistance or a significant reduction in hours worked or a sheltered situation or was on sick leave 1
- has not been working or studying in any capacity and not managing own household 0

DAILY LIVING   *During the last week, the patient*

- has been self-reliant in eating, washing, toileting and dressing; using public transport or driving own car 2
- has been requiring assistance (another person or special equipment) for daily activities and transport but performing light tasks 1
- has not been managing personal care or light tasks and/or not leaving own home or institution at all 0

☐

301

Table 10.2. (continued)

HEALTH

During the last week, the patient
- has been appearing to feel well or reporting feeling 'great' most of the time 2
- has been lacking energy or not feeling entirely 'up to par' more than just occasionally 1
- has been feeling very ill or 'lousy', seeming weak and washed out most of the time, or was unconscious 0

☐

SUPPORT

During the last week
- the patient has been having good relations with others and receiving strong support from at least one family member and/or friend 2
- support received or perceived has been limited from family and friends and/or by the patient's condition 1
- support from family and friends occurred infrequently or only when absolutely necessary or patient was unconscious 0

☐

OUTLOOK

During the past week the patient
- has usually been appearing calm and positive in outlook, accepting and in control of personal circumstances, including surroundings 2
- has sometimes been troubled because not fully in control of personal circumstances or has been having periods of obvious anxiety or depression 1
- has been seriously confused or very frightened or consistently anxious and depressed or unconscious 0

☐

QL INDEX TOTAL ☐☐

How confident are you that your scoring of the preceding dimensions is accurate? Please ring the appropriate category.

Absolutely Confident	Very Confident	Quite Confident	Not Very Confident	Very Doubtful	Not at all Confident
1	2	3	4	5	6

☐

Table 10.3. *Scales and Definitions for the Classification of Function Levels*

Scale	Step	Definition
Mobility scale		
5	Travelled freely	Used public transportation or drove alone. For below 6 age group, travelled as usual for age.
4	Travelled with difficulty	(a) Went outside alone, but had trouble getting around community freely, or (b) required assistance to use public transportation or automobile.
3	In house	(a) All day, because of illness or condition, or (b) needed human assistance to go outside.
2	In hospital	Not only general hospital, but also nursing home, extended care facility, sanatorium, or similar institution.
1	In special unit	For some part of the day in a restricted area of the hospital such as intensive care, operating room, recovery room, isolation ward, or similar unit.
0	Death	
Physical activity scale		
4	Walked freely	With no limitations of any kind.
3	Walked with limitations	(a) With cane, crutches, or mechanical aid, or (b) limited in lifting, stooping, or using stairs or inclines, or (c) limited in speed or distance by general physical condition.
2	Moved independently in wheelchair	Propelled self alone in wheelchair.
1	In bed or chair	For most or all of the day.
0	Death	
Social activity scale		
5	Performed major and other activities	*Major* means specifically: play for below 6, school for 6–17, and work or maintain household for adults. *Other* means all activities not classified as major, such as athletics, clubs, shopping,

Table 10.3. (*continued*)

Scale	Step	Definition
		church, hobbies, civic projects, or games as appropriate for age.
4	Performed major but limited in other activities	Played, went to school, worked, or kept house but limited in other activities as defined above.
3	Performed major activity with limitation	Limited in the amount or kind of major activity performed, for instance, needed special rest periods, special school, or special working aids.
2	Did not perform major activity but performed self-care activities	Did not play, go to school, work or keep house, but dressed, bathed, and fed self.
1	Required assistance with self-care activities	Required human help with one or more of the following – dressing, bathing, or eating – and did not perform major or other activities. For below 6 age group, means assistance not usually required for age.
0	Death	

have in a particular life can be adjusted for by making the measure relative to what had been the individual's normal level of activity in each of these areas prior to sickness. The two broad groups of functions, physical and psychosocial, are each broken down into several distinct components. For each primary function, the SIP measures the impact of sickness by eliciting information concerning whether activities typical in the exercise of that function continue to be performed, or have become limited. Even for primary functions, about which it is plausible to claim that they have a place in virtually any life, the different functions can have a different *relative* value or importance within different lives, and the SIP makes no attempt to measure those differences. The QLI likewise addresses a person's levels of activity in daily living, specifically measuring the presence of related behaviours in the relevant areas. In measuring health and outlook, the primary concern is with subjective feeling states of the person, though here too there is concern with relevant behaviour. The category of support addresses both the social behaviour of the individual and the availability of people in the individual's environment to provide such relationships. This category illustrates the important point that most primary functional capacities require both behavioural capacities in the individual and relevant resources in the individual's external environment. The HSI addresses three broad categories of primary function – mobility, physical activity, and social activity – with evidence of current functional capacity found in current levels of activity. It is noteworthy that even this index, which focuses explicitly on the health status of individuals, does not measure the presence or absence of disease, as one might expect given common understandings of 'health' as the absence of disease; like the SIP and QLI, it too measures levels of very broad primary functions.

A second important feature of these measures shows up explicitly only in the first two – SIP and QLI – and is best displayed in the 'outlook' category of the QLI, though it is also at least partly captured in the 'emotional behaviour' category of the SIP. Both these categories can be understood as attempts to capture people's subjective response to their objective physical condition and level of function, or, in short, their level of happiness or satisfaction with their lives, though the actual measures are far too crude to measure happiness with much sensitivity. The important point is that the use of these categories represents a recognition that *part* of what makes a good life is that the person in question is happy or pleased with how it is going; that is, subjectively experiences it as going well, as fulfilling his or her major aims, and as satisfying. This subjective happiness component is not unrelated, of course, to how well the person's life is going as measured by the level of the other objective primary functions. How happy we are with our lives is significantly determined by how well our lives are in fact going in other objective respects. Nevertheless, medicine provides many examples that show it is a mistake to assume that the subjective happiness component correlates closely and invariably with other objective functional measures. In one study, for example, researchers found a substantial relation between different objective function variables and also between different subjective response or outlook variables, but only a very limited relation between objective and subjective variables.[46] These data reinforce the importance of including both objective function and subjective response categories in a full conception of the quality of life, since neither is a reliable surrogate for the other. Given this at least partial independence between happiness and function variables, what is their relative weight in an overall assessment of a good life? Here, too, medicine brings out forcefully that there can be no uniform answer to this question. In the face of seriously debilitating injuries, one patient will adjust her aspirations and expectations to her newly limited functional capacities and place great value on achieving happiness despite these limitations. Faced with similar debilitating injuries, another patient will assign little value to adjusting to the disabilities in order to achieve happiness in spite of them, stating that she 'does not want to become the kind of person who is happy in that debilitated and dependent state'.[47]

There are other important qualifications of the generally positive relation between this happiness component of a good life and both the other primary function components and the overall assessment of how good a life it is. These qualifications are not all special to health care and the quality of life, but some are perhaps more evident and important in the area of health care than elsewhere. The first qualification concerns people's adjustments to limitations of the other primary functions. Sometimes the limitation in function, or potential limitation, is due to congenital abnormalities or other handicaps present from birth. For example, an American television program recently reported on a follow-up of some of the children, now young adults, born to pregnant women who had taken the drug Thalidomide in the late 1950s.[48] The people reported on had suffered no brain damage but had been born with serious physical deformities, including lacking some or any arms and legs. While this placed many impairments in the way of carrying out primary functions such as eating, working, home management, physical mobility, and ambulation in the manner of normal adults, these people had made remarkable adjustments to compensate for their physical limitations: one was able to perform all the normal functions of eating using his foot in place of missing arms and hands; another made his living as an artist painting with a brush held between his teeth; another without legs was able to drive in a specially equipped car; and a mother of three without legs had adapted so as to be able to perform virtually all the normal tasks of managing a family and home.

These were cases where physical limitations that commonly restrict and impair people's primary functional capacities and overall quality of life had been so well compensated for as to enable them to perform the *same* primary functions, though in different ways, as well as normal, unimpaired persons do. While a few life plans possible for others remained impossible for them because of their limitations (for example, being professional athletes), their essentially unimpaired level of primary functions as a result of the compensations they had made left them with choice from among a sufficiently wide array of life plans that it is probably a mistake to believe that their quality of life had been lowered much or at all by their impairments. These cases illustrate that even serious physical limitations do not always lower quality of life if the disabled persons have been able or helped sufficiently to compensate for their disabilities so that their level of primary functional capacity remains essentially unimpaired; in such cases it becomes problematic even to characterize those affected as disabled.

In other cases, compensating for functional disabilities, particularly when they arise later in life, may require adjustments involving substantial changes in the kind of work performed, social and recreational activities pursued, and so forth. When these disabilities significantly restrict the activities that had been and would otherwise have been available to and pursued by the person, they will, all other things being equal, constitute reductions in the person's quality of life. If they do so, however, it will be because they significantly restrict the choices, or what Norman Daniels has called the normal opportunity range, available to the persons, and not because the compensating paths chosen need be, once entered on, any less desirable or satisfying.[49] The *opportunity for choice* from among a reasonable array of life plans is an important and independent component of quality of life: it is insufficient to measure only the quality of the life plan the disabled person now pursues and his or her satisfaction with it. Adjustments to impairments that leave primary functions undiminished or that redirect one's life plan into areas where function will be better – both central aims of rehabilitative medicine – can, however, enhance quality of life even in the face of a diminished opportunity range.

In his theory of just health care Daniels uses the notion of an *age-adjusted* normal opportunity range, which is important for the relation between opportunity and quality of life or a good life. Some impairments in primary functions occur as common features of even the normal aging process, for example, limitations in previous levels of physical activity. Choosing to adjust the nature and level of our planned activities to such impairments in function is usually considered a healthy adjustment to the aging process. This adjustment can substantially diminish the reduction in the person's quality of life from the limitations of normal aging. Nevertheless, even under the best of circumstances, the normal aging process (especially, say, beyond the age of 80) does produce limitations in primary functions that will reduce quality of life. Thus, while quality of life must always be measured against normal, primary functional capacities for humans, it can be diminished by reductions both in individual function below the age-adjusted norm and by reductions in normal function for humans as they age.

I have suggested above that adjustments in chosen pursuits as a result of impairments in primary, or previously pursued individual, functions can compensate substantially (fully, in the effects on happiness) for impaired function, but will often not compensate fully for significant reductions in the range of *opportunities* available for choice, and so will not leave quality of life undiminished. In some cases, however, a patient's response and adjustment to the limitations of illness or injury may be so complete, as regards his commitment to and happiness from the new chosen life path, that there is reason to hold that his quality of life is as high as before, particularly as he gets further away in time from the onset of the limiting illness or injury and as the new life becomes more securely and authentically the person's own. An undiminished or even increased level of happiness and satisfaction, together with an increased commitment to the new life, often seem the primary relevant factors when they are present. But we must also distinguish different *reasons* why the affective or subjective component of quality of life, which I have lumped under the notion of happiness, may remain undiminished, since this is important for an evaluation of the effect on quality of life.

A person's happiness is to some significant extent a function of the *degree* to which his or her major aims are being at least reasonably successfully pursued. Serious illness or injury resulting in serious functional impairment often requires a major revaluation of one's plan of life and its major aims and expectations. Over time, such revaluations can result in undiminished or even increased levels of happiness, despite decreased function, because the person's aspirations and expectations have likewise been revised and reduced. The common cases in medicine in which, following serious illness, people come to be satisfied with much less in the way of hopes and accomplishments illustrate clearly the incompleteness of happiness as a full account of the quality of life. To be satisfied or happy with getting much less from life, because one has come to expect much less, is still to get *less* from life or to have a less good life. (The converse of this effect is when rising levels of affluence and of other objective primary functions in periods of economic development lead to even more rapidly rising aspirations and expectations, and in turn to an *increasing gap* between accomplishments and expectations.) Moreover, whether the relation of the person's choices to his aspirations and expectations reflects his exercise of self-determination in response to changed circumstances is important in an overall assessment of his quality of life and shows another aspect of the importance of self-determination to quality of life.

Illness and injury resulting in serious limitations of primary functions often strike individuals without warning and seemingly at random, and are then seen by them and others as a piece of bad luck or misfortune. Every life is ended by death, and few people reach death after a normal life span without some serious illness and attendant decline in function. This is simply an inevitable part of the human condition. Individual character strengths and social support services enable people unfortunately impaired by disease or injury to adjust their aims and expectations realistically to their adversity, and then to get on with their lives, instead of responding to their misfortune with despair and self-pity. Circumstances beyond individuals' control may have dealt them a cruel blow, but they can retain dignity as self-determining agents capable of responsible choice in directing and retaining control over their lives within the limits that their new circumstances permit. We generally admire people who make the best of their lot in this way, and achieve happiness and accomplishment despite what seems a cruel fate. This reduction in aims and expectations, with its resultant reduction in the gap between accomplishments and aims, and the in turn resultant increase in happiness, is an outcome of the continued exercise of self-determination. It constitutes an increase in the happiness and self-determination components of quality of life though, of course, only in response to an earlier decrease in the person's level of primary functions.

Other ways of reducing this gap between accomplishments and expectations bypass the person's self-determination and are more problematic as regards their desirability and their effect on a person's quality of life. Jon Elster has written, for example, outside of the medical context, of different kinds of non-autonomous preferences and preference change.[50]

Precisely characterizing the difference between what Elster calls non-autonomous preferences and what I have called the exercise of self-determination in adjusting to the impact of illness and injury raises deep and difficult issues that I cannot pursue here. Nevertheless, I believe that response to illness through the exercise of self-determined choice, in the service of protecting or restoring quality of life, is one of the most important practical examples of the significance for overall assessments of the quality of life of *how* to achieve the reasonable accord between aims and accomplishments that happiness requires.

CONCLUSION

Let us tie together some of the main themes in accounts of the quality of life or of a good life suggested by the literature in medical ethics and health policy. While that literature provides little in the way of well-developed, philosophically sophisticated accounts of the quality of life or of a good life, it is a rich body of analysis, data, and experience on which philosophical accounts of a good life can draw. I have presented here at least the main outlines of a general account of a good life suggested by that work. The account will be a complex one which, among the main philosophical theories distinguished earlier, probably most comfortably fits within ideal theories. I have suggested that we can employ Sen's construction of a plurality of independent vectors, each of which is an independent component of a full assessment of the degree to which a person has a good life.

The ethical frameworks for medical treatment decision-making bring out the centrality of a person's capacity as a valuing agent, or what I have called self-determination, in a good life. The capacity for and exercise of self-determination can be taken to be a – or I believe *the* – fundamental ideal of the person within medical ethics. The exercise of self-determination in constructing a relatively full human life will require in an individual four broad types of primary functions: biological, including, for example, well-functioning organs; physical, including, for example, ambulation; social, including, for example, capacities to communicate; mental, including, for example, a variety of reasoning and emotional capabilities. There are no sharp boundaries between these broad types of primary function, and for different purposes they can be specified in more or less detail and in a variety of different bundles. The idea is to pick out human functions that are necessary for, or at least valuable in, the pursuit of nearly all relatively full and complete human life plans. These different functions can be represented on different vectors and they will be normatively objective components of a good life, though their relative weight within any particular life may be subjectively determined.

There are in turn what we can call agent-specific functions, again specifiable at varying levels of generality or detail, which are necessary for a person to pursue successfully the particular purposes and life plan he or she has chosen: examples are functional capacities to do highly abstract reasoning of the sort required in mathematics or philosophy and the physical dexterity needed for success as a musician, surgeon, or athlete. Once again, these functions can be represented on independent vectors, though their place in the good life for a particular person is determined on more normatively subjective grounds depending on the particular life plan chosen. The relative weight assigned to agent-specific functions and, to a substantially lesser degree, to primary functions, will ultimately be determined by the valuations of the self-determining agent, together with factual determinations of what functions are necessary in the pursuit of different specific life plans. The centrality of the valuing and choosing agent in this account of a good life gives both primary and, to a lesser extent, agent-specific functional capacities a central place in the good life because of their necessary role in making possible a significant range of opportunities and alternatives for choice.

At a more agent-specific level still are the particular desires pursued by people on particular occasions in the course of pursuing their valued aims and activities. Different desires and the degree to which they can be successfully satisfied can also be represented using the vector approach. It bears repeating that the level of a person's primary functional capacities, agent-specific functional capacities, and satisfaction of specific desires will all depend both on properties of the agent and on features of his or her environment that affect those functional capacities and desire pursuits. The inclusion of primary functions, agent-specific functions, and the satisfaction of specific desires all within an account of the good life allows us to recognize both its normatively objective and normatively subjective components. Analogously, these various components show why we can expect partial, but only partial, interpersonal comparability of the quality of life or of good lives – comparability will require interpersonal overlapping of similarly weighted primary functions, agent-specific functions, and specific desires. The importance of functional capacities at these different levels of generality reflects the centrality of personal choice in a good life and the necessity for a choice of alternatives and opportunities.

Finally, there will be the hedonic or happiness component of a good life, that aspect which represents a person's subjective, conscious response in terms of enjoyments and satisfactions to the life he or she has chosen and the activities and achievements it contains. These may be representable on a single vector or on a number of distinct vectors if the person has distinct and incommensurable satisfactions and enjoyments. Happiness will usually be only partially dependent on the person's relative success in satisfying his or her desires and broader aims and projects. Once again, it is the valuations of the specific person in question that will determine the relative weight the happiness vector receives in the overall account of a good life for that person.

Needless to say, in drawing together these features of an account of the quality of life or of a good life from the medical ethics and health policy literatures, I have done no more than sketch a few of the barest bones of a full account of a good life. However, even these few bones suggest the need for more complex accounts of the quality of life than are often employed in programs designed to improve the quality of life of real people.

NOTES

1 Breslow, 1972: 347–55.
2 Parfit, 1984, cf. esp. app. 1.
3 See T. M. Scanlon's discussion (in Sen and Nussbaum, 1992) of these alternative theories. What I call preference satisfaction and ideal theories he calls desire and substantive good theories.
4 Some time ago I discussed these as alternative interpretations of utility (Brock, 1973). The most subtle and detailed recent discussion of these alternative theories is Griffin, 1986, chs. 1–4.
5 Virtually all discussions of desire or preference satisfaction theories of the good contain some provision for correcting preferences. One of the better treatments, with extensive references to the literature, is Goodin, 1986: 75–101.
6 What I call ideal theories are what Parfit (1984) calls 'objective list' theories. I prefer the label 'ideal theories', because what is usually distinctive about this kind of theory is its proposal of specific, normative ideals of the person.
7 See the papers by H. Putnam, R. A. Putnam, and M. Walzer, in Sen and Nussbaum, 1992.
8 Morreim, 1986: 45–69.
9 Sen's main discussion of the 'vector view' applied to the notion of utility is in Sen, 1980. I am much indebted to Sen's subtle discussions in a number of places of distinctions of importance to conceptions of the quality of life and of a good life. See esp. Sen, 1985*a*, 1985*b*, 1987, and his paper in Sen and Nussbaum, 1992
10 Kass, 1985.
11 Ibid., 171.
12 Ibid., 173. For a more philosophically sophisticated analysis of the concept of health that also construes it in functional terms as a natural, biological norm not involving value judgements, see Boorse, 1975, 1977. One of the most useful collections of papers on concepts of health is Caplan, Engelhardt, and McCartney, 1981.
13 I make no attempt here to provide any more than a few rep-

resentative references to this very large literature. Probably the single best source for the medical ethics literature in this area is the *Hastings Center Report*. In the medical literature, see Wanzer, 1984, and Ruark, 1988. For a good review of most of the principal legal decisions in the United States concerning life-sustaining treatment, see Annas and Glantz, 1986. The most influential treatment of these issues by a governmental body in the United States is the report of the President's Commission for the Study of Ethical Problems in Medicine and Biomedical and Behavioral Research (1983). See also the recent report by the Hastings Center (1987). For discussions of Living Wills and Durable Powers of Attorney see Steinbrook and Lo, 1984, and Schneiderman and Arras, 1985. An application to clinical practice of the consensus that patients should have rights to decide about their care is Jonsen, Siegler, and Winslade, 1982.

14 An influential statement of the shared decision-making view is another report of the President's Commission for the Study of Ethical Problems in Medicine and Biomedical and Behavioral Research (1982). A sensitive discussion of the difficulties of achieving shared decision-making in clinical practice is Katz, 1984.

15 E.g. Ramsey, 1978: 206–7.

16 I have discussed some of the ethical implications of different measures for valuing lives found in the economic and policy literatures (Brock, 1986).

17 It has been argued that this is the proper understanding of the distinction between 'ordinary' and 'extraordinary' treatment. That is, extraordinary treatment is treatment that for the patient in question and in the circumstances that obtain is unduly burdensome. Cf. President's Commission, 1983: 82–9.

18 Rango, 1985.

19 Ibid., 836, quoting E. M. Brody.

20 The importance of this factor was stressed in a widely publicized legal decision concerning the use of painful chemotherapy for a man suffering from cancer who had been severely retarded from birth (*Superintendent of Belchertown State School v. Saikewicz*, 1977).

21 Rango, 1985: 838.

22 McCormick, 1974.

23 Rhoden, 1985. Another sensitive discussion of the need for

quality of life judgments in treatment decisions for imperilled new-born babies is Arras, 1984.

24 The most comprehensive treatment of the informed consent doctrine is Faden and Beauchamp, 1986. See also President's Commission, 1982. The exceptions to the legal requirement of informed consent are discussed in Meisel, 1979.

25 When patients' settled preferences are not in accord with their values, then their informed and voluntary choices may not reflect their conception of their good. One of the clearest examples is the patient who is addicted to morphine, hates his addiction and tries unsuccessfully to resist it, but in the end is overpowered by his desire for the morphine and takes it. This in essence is Frankfurt's example in his classic paper (Frankfurt, 1971). Frankfurt's analysis is in terms of first- and second-order desires, but can also be put in terms of the desires a person in fact has as opposed to the desires the person values and wants to have. When these are in conflict his informed and voluntary choice may not reflect the values that define his own conception of his good. There is a sense in which his choice in these conditions is involuntary, so it would be possible to extend the informed consent doctrine's requirement of voluntariness to include this sense. Alternatively, it might be possible to interpret the requirement of competence in a way that makes the morphine addict incompetent to decide whether to continue using morphine.

26 This account of the principal values underlying the informed consent doctrine as patient well-being and self-determination is common to many analyses of that doctrine; cf. President's Commission, 1982. I have employed it in Brock, 1987.

27 *Schloendorff* v. *Society of New York Hospital*, 1914.

28 Cf. Faden and Beauchamp, 1986: ch. 4.

29 *Natanson* v. *Kline*, 1960.

30 Mill, 1859: 13.

31 The most detailed recent account of justified paternalism in the spirit of Mill's position is VanDeVeer, 1986. The most sophisticated development of a Millian position on paternalism in criminal law is Feinberg, 1986. I have explored some of the issues between rights-based and consequentialist accounts of paternalism in Brock, 1983 and 1988.

32 This conventional view is reflected in the independent ethical principles of beneficence and autonomy in Beauchamp and

Childress, 1979. This book has probably been the account of moral principles most influential with people in medicine and health care without philosophical training in ethics.

33 Cf. Rawls, 1980, and Sen, 1985, esp. pp. 203–4.

34 If personal well-being is understood in this way, it suggests that satisfaction of a person's non-personal desires that make no such reference to him do not increase the person's well-being. Consider a loyal fan of the Boston Red Sox baseball team, who wants the Red Sox to win the pennant. On the last day of the season, tied for first place with the New York Yankees, the Red Sox beat the Yankees and win the pennant. Suppose the fan is travelling in a remote area of Alaska on the day of the big game and a week later, before getting out of the wilderness area and learning of the Red Sox victory, he is killed by a rock slide. Was his personal well-being increased at all simply because the state of affairs he desired – the Red Sox victory – obtained? I believe the answer should be no.

The harder question is whether, in our broader sense of a good life, he had a better life even unbeknownst to him. And was the quality of his life any better? Certainly his life *as experienced by him* was no better and not of higher quality. Even in our broad sense of a good life, his life may seem not to have gone better, but perhaps that is only because this is a relatively unimportant desire. Suppose instead, to adapt an example of Parfit's (1984), a person devotes fifty years of his life to saving Venice and then, confident that it is safe, goes on vacation to the Alaskan wilderness. While he is there a flood destroys Venice, but, like the Red Sox fan, he never learns of it because a week after the flood and before getting out of the wilderness he is killed by a rock slide. Parfit notes that it is plausible to say of the destruction of Venice both that it has made the person's life go less well because he had invested his life in this goal and his life's work is now in vain, but also that it cannot lower the quality of his life if it does not affect the quality of his experience. This suggests a point where a broad notion of a good life may diverge from the notion of the quality of life. Since I believe that medicine and medical ethics have little illumination to offer on this point, I set it aside here and shall in the body of the paper continue to use the broad notion of a good life largely interchangeably with the quality of life.

35 Allen Buchanan and I have discussed ethical issues in decision-

making for incompetent persons in our paper (Buchanan and Brock, 1986: 17–94) and in our book (Buchanan and Brock, 1989).

36 Rebecca Dresser has developed a Parfitian challenge to the substituted judgment principle as well as to the authority of advance directives in cases in which the conditions creating the patient's incompetence also reduce or eliminate the psychological continuity and connectedness necessary for personal identity to be maintained. Cf. Dresser, 1986. Cf. also Buchanan, 1988.

37 Scanion argues (in Sen and Nussbaum, 1992) that desire satisfaction is not itself a basic part of individual well being, but is dependent on hedonistic or ideal (what he calls substantive good) reasons for its support.

38 Two of the more important discussions of life plans and of how they can give structure and coherence to life are Fried, 1970: ch. 10, and Rawls, 1971: ch. 7.

39 The notion of a biographical life is employed in the medical ethics literature by Rachels, 1986: 5–6, by Singer and Kuhse, 1985: 129–39, and by Callahan, 1987. It is also implicit in Tooley's (1983) account of the right to life.

40 Among the useful papers consulted for this third section on health policy measures of quality of life, and not cited in other notes, are: Anderson 1986; Berg, 1986; Bergner *et al.*, 1976; Calman, 1984; Cohen, 1982; Cribb, 1985; Editorial, 1986; Edlund and Tancredi; 1985; Flanagan, 1982; Gehrmann, 1978; Gillingham and Reece, 1980; Grogono, 1971; Guyatt, 1986; Hunt and McEwen, 1980; Katz, 1963; Kornfeld, 1982; Klotkem, 1982; Liang, 1982; Najman and Levine, 1981; Pearlman and Speer, 1983; Presant, 1984; Report, 1984; Starr, 1986; Sullivan, 1966; Thomasma, 1984, 1986; Torrance, 1972, 1976*a*, 1976*b*.

41 An example of a broad quality of life measure not focused on health care and disease can be found in the Swedish Level of Living Surveys discussed in the essay by Erikson in Sen and Nussbaum, 1992.

42 Bergner *et al.*, 1981.

43 Spitzer *et al.*, 1981.

44 Chen, Bush, and Patrick, 1975.

45 Rawls, 1971: 62, 90–5.

46 Evans, 1985.

47 The main character in the popular play and subsequent film

Whose Life is it Anyway?, having become paralysed from the neck down, displayed this attitude of not wanting to become a person who had adjusted to his condition.
48 *60 Minutes*, CBS television network program, 21 February 1988.
49 Daniels, 1985: chs. 2 and 3.
50 Elster, 1982.

BIBLIOGRAPHY

Anderson, John P., *et al.* (1986). 'Classifying Function for Health Outcome and Quality-of-Life Evaluation', *Medical Care*, 24, 454–71.

Annas, George, and Glantz, Leonard (1986). 'The Right of Elderly Patients to Refuse Life-Sustaining Treatment'. *Milbank Quarterly*, vol. 64, suppl. 2, 95–162.

Arras, John (1984). 'Toward an Ethic of Ambiguity', *Hastings Center Report*, 14(Apr.), 25–33.

Beauchamp, Tom L., and Childress, James F. (1979). *Principles of Biomedical Ethics*. New York: Oxford University Press.

Berg, Robert L. (1986). 'Neglected Aspects of the Quality of Life', *Health Services Research*, 21, 391–5.

Bergner, Marilyn, *et al.* (1976). 'The Sickness Impact Profile: Conceptual Formulation and Methodology for the Development of a Health Status Measure', *International Journal of Health Services*, 6, 393–415.

—— (1981). 'The Sickness Impact Profile: Development and Final Revision of a Health Status Measure', *Medical Care*, 19, 787–805.

Boorse, Christopher (1975). 'On the Distinction between Disease and Illness', *Philosophy and Public Affairs*, 5, 49–68.

—— (1977). 'Health as a Theoretical Concept', *Philosophy of Science*, 44.

Breslow, Lester (1972). 'A Quantitative Approach to the World Health Organization Definition of Health: Physical, Mental and Social Well-being', *International Journal of Epidemiology*, 1, 347–55.

Brock, Dan W. (1973). 'Recent Work in Utilitarianism', *American Philosophical Quarterly*, 10, 241–76.

—— (1983). 'Paternalism and Promoting the Good', in Rolf Sartorius (ed.), *Paternalism*. Minneapolis: University of Minnesota Press.

—— (1986). 'The Value of Prolonging Human Life', *Philosophical Studies*, 50, 401–28.

—— (1987). 'Informed Consent', in Tom Regan and Donald VanDeVeer (eds.), *Health Care Ethics*. Philadelphia: Temple University Press.

—— (1988). 'Paternalism and Autonomy', *Ethics*, 98, 550–65.

Buchanan, Allen (1988) 'Advance Directives and the Personal Identity Problem', *Philosophy and Public Affairs* 17, 277–302.

—— and Brock, Dan W. (1986). 'Deciding for Others', *Milbank Quarterly*, vol. 64, suppl. 2, 17–94.

—— —— (1989). *Deciding For Others*. Cambridge: Cambridge University Press.

Callahan, Daniel (1987). *Setting Limits*. New York: Simon and Schuster.

Calman, K. C. (1984). 'Quality of Life in Cancer Patients: An Hypothesis', *Journal of Medical Ethics*, 10, 124–7.

Caplan, Arthur L., *et al.* (1981). *Concepts of Health and Disease: Interdisciplinary Perspectives*. Reading, Mass.: Addison-Wesley.

Chen, Milton M., Bush, J. W., and Patrick, Donald L. (1975). 'Social Indicators for Health Planning and Policy Analysis', *Policy Sciences*, 6, 71–89.

Cohen, Carl (1982). 'On the Quality of Life: Some Philosophical Reflections', *Circulation*, vol. 66 suppl. 3, 29–33.

Cribb, Alan (1985). 'Quality of Life: A Response to K. C. Calman'. *Journal of Medical Ethics*, 11, 142–5.

Daniels, Norman (1985). *Just Health Care*. Cambridge: Cambridge University Press.

Dresser, Rebecca (1986). 'Life, Death, and Incompetent Patients: Conceptual Infirmities and Hidden Values in the Law', *Arizona Law Review*, 28, 373–405.

Editorial (1986). 'Assessment of Quality of Life in Clinical Trials'. *Acta Medica Scandia*, 220, 1–3.

Edlund, Mathew, and Tancredi, Lawrence (1985). 'Quality of Life: An Ideological Critique', *Perspectives in Biology and Medicine*, 28, 591–607.

Elster, Jon (1982). 'Sour Grapes: Utilitarianism and the Genesis of Wants', in Amartya Sen and Bernard Williams (eds.), *Utilitarianism and Beyond*. Cambridge: Cambridge University Press.

Evans, Roger W. (1985). 'The Quality of Life of Patients with End Stage Renal Disease', *New England Journal of Medicine*, 312, 553–9.

Faden, Ruth R., and Beauchamp, Tom L. (1986). *A History and Theory of Informed Consent*. New York: Oxford University Press.

Feinberg, Joel. (1986). *Harm to Self*. New York: Oxford University Press.

Flanagan, John C. (1982). 'Measurement of Quality of Life: Current State of the Art'. *Archives of Physical Rehabilitation Medicine*, 63, 56–9.

Frankfurt, Harry (1971). 'Freedom of the Will and the Concept of a Person', *Journal of Philosophy*, 68, 5–20.

Fried, Charles (1970). *An Anatomy of Values*. Cambridge, Mass.: Harvard University Press.

Gehrmann, Friedhelm (1978). '"Valid" Empirical Measurement of Quality of Life', *Social Indicators Research*, 5, 73–109.

Gillingham, Robert, and Reece, William S. (1980). 'Analytical Problems in the Measurement of the Quality of Life', *Social Indicators Research*, 7, 91–101.

Goodin, Robert (1986). 'Laundering Preferences', in Jon Elster and Aanund Hylland (eds.), *Foundations of Social Choice Theory*. Cambridge: Cambridge University Press.

Griffin, James (1986). *Well-Being*. Oxford: Oxford University Press.

Grogono, A. W. (1971). 'Index for Measuring Health', *Lancet*, vol. 2 for 1971, 1024–6.

Guyatt, Gordon H., *et al.* (1986). 'Measuring Disease-Specific Quality of Life in Clinical Trials', *Canadian Medical Association Journal*, 134, 889–95.

Hastings Center (1987). *Guidelines on the Termination of Treatment and the Care of the Dying*. Briarcliff Manor, NY: The Hastings Center.

Hunt, Sonya, and McEwen, James (1980). 'The Development of a Subjective Health Indicator', *Sociology of Health and Illness*, 2, 203–31.

Jonsen, Albert R., Siegler, Mark, and Winslade, William J. (1982). *Clinical Ethics*. New York: Macmillan.

Kass, Leon (1985). *Toward a More Natural Science*. New York: Free Press.

Katz, Jay (1984). *The Silent World of Doctor and Patient*. New York: Free Press.

Katz, Sidney, *et al.* (1963). 'Studies of Illness in the Aged. The Index of ADL: A Standardized Measure of Biological and Psychosocial Function', *JAMA*, 185, 914–19.

Klotkem, Frederic J. (1982). 'Philosophic Considerations of Quality of Life for the Disabled', *Archives of Physical Rehabilitation Medicine*, 63, 59–63.

Kornfeld, Donald S., *et al.* (1982). 'Psychological and Behavioral Responses After Coronary Artery Bypass Surgery', *Circulation*, vol. 66, suppl. 3, 24–8.

Liang, Mathew, *et al.* (1982). 'In Search of a More Perfect Mousetrap (Health Status or Quality of Life Instrument)', *Journal of Rheumatology*, 9, 775–9.

McCormick, S.J., Richard J. (1974). 'To Save or Let Die: The Dilemma of Modern Medicine', *JAMA*, 229, 172–6.

Meisel, Alan (1979). 'The "Exceptions" to the Informed Consent Doctrine: Striking a Balance Between Competing Values in Medical Decision-making', *Wisconsin Law Review*, 413–88.

Mill, J. S. (1859). *On Liberty*. Indiana/New York: Bobbs-Merrill, 1956.

Morreim, E. Haavi (1986). 'Computing the Quality of Life', in G. J. Agich and C. E. Begley (eds.), *The Price of Health*. Dordrecht: D. Reidel.

Najman, Jackob, and Levine, Sol (1981). 'Evaluating the Impact of Medical Care and Technologies on the Quality of Life: A Review and Critique', *Social Science and Medicine*, 15F, 107–15.

Natanson v. Kline (1960) 186 Kan. 393, 350 p. 2d 1093.

Parfit, Derek (1984). *Reason and Persons*. Oxford: Oxford University Press.

Pearlman, Robert, and Speer, James (1983). 'Quality of Life Considerations in Geriatric Care', *Journal of the American Geriatrics Society*, 31, 113–20.

Presant, Cary A. (1984). 'Quality of Life in Cancer Patients', *American Journal of Clinical Oncology*, 7, 571–3.

President's Commission for the Study of Ethical Problems in Medicine and Biomedical and Behavioral Research (1982). *Making Health Care Decisions*. Washington: US Government Printing Office.

––––– (1983). *Deciding to Forgo Life-Sustaining Treatment*. Washington: US Government Printing Office.

Rachels, James (1986). *The End of Life*. Oxford: Oxford University Press.

Ramsey, Paul (1978). *Ethics at the Edge of Life*. New Haven, Conn.: Yale University Press.

Rango, Nicholas (1985). 'The Nursing Home Resident with Dementia', *Annals of Internal Medicine*, 102, 835–41.

Rawls, John (1971). *A Theory of Justice.* Cambridge, Mass.: Harvard University of Press.

—— (1980). 'Kantian Constructivism in Moral Theory', *Journal of Philosophy*, 77, 515–72.

Report (1984). 'The 1984 Report of the Joint National Committee on Detection, Evaluation, and Treatment of High Blood Pressure', *Archives of Internal Medicine*, 144, 1045–57.

Rhoden, Nancy K. (1985). 'Treatment Dilemmas for Imperiled Newborns: Why Quality of Life Counts', *Southern California Law Review*, 58, 1283–1347.

Ruark, John E., *et al.* (1988). 'Initiating and Withdrawing Life Support', *New England Journal of Medicine* 318 (Jan.), 25–30.

Schloendorf v. *Society of New York Hospital* (1914). 211 N.Y. 125, 105 N.E. 92, 95.

Schneiderman, Lawrence, and Arras, John (1985). 'Counseling Patients to Counsel Physicians on Future Care in the Event of Patients' Incompetence', *Annals of Internal Medicine*, 102, 693–8.

Sen, Amartya (1980). 'Plural Utility', *Proceedings of the Aristotelian Society*, 81, 193–218.

—— (1985*a*). *Commodities and Capabilities.* Amsterdam: Elsevier Science Publishers.

—— (1985*b*). 'Well-being, Agency and Freedom: The Dewey Lectures', *Journal of Philosophy*, 82, 169–221.

—— (1987). *The Standard of Living.* Cambridge: Cambridge University Press.

Sen, A., and Nussbaum, M. (1992). *The Quality of Life.* Oxford: Oxford University Press.

Singer, Peter, and Kuhse, Helga (1985). *Should This Baby Live?* Oxford: Oxford University Press.

Spitzer, Walter O., *et al.* (1981). 'Measuring the Quality of Life of Cancer Patients: A Concise QL-Index for Use by Physicians', *Journal of Chronic Disease*, 34, 585–97.

Starr, T. Jolene, *et al.* (1986). 'Quality of Life and Resuscitation Decisions in Elderly Patients', *Journal of General Internal Medicine*, 1, 373–9.

Steinbrook, Robert, and Lo, Bernard (1984). 'Decision-making for Incompetent Patients by Designated Proxy: California's New Law', *New England Journal of Medicine*, 310, 1598–1601.

Sullivan, Daniel F. (1966). 'Conceptual Problems in Developing an Index of Health', *Vital and Health Statistics*, 2, 1–18.

Superintendent of Belchertown State School v. *Saikewicz* (1977). 370 N.E. 2d 417.

Thomasma, David C. (1984). 'Ethical Judgment of Quality of Life in the Care of the Aged', *Journal of the American Geriatrics Society*, 32, 525–7.

—— (1986). 'Quality of Life, Treatment Decisions, and Medical Ethics', *Clinics in Geriatric Medicine*, 2, 17–27.

Tooley, Michael. (1983). *Abortion and Infanticide*. Oxford: Oxford University Press.

Torrance, George W. (1972). 'Social Preferences for Health States: An Empirical Evaluation of Three Measurement Techniques', *Socio-Economic Planning Sciences*, 10, 129–36.

—— (1976a). 'Toward a Utility Theory Foundation for Health Status Index Models', *Health Services Research*, 10, 129–36.

—— *et al.* (1976b). 'A Utility Maximization Model for Evaluation of Health Care Programs', *Health Services Research*, 6, 118–33.

Van DeVeer, Donald (1986). *Paternalistic Intervention*. Princeton, NJ: Princeton University Press.

Wanzer, Sidney H., *et al.* (1984). 'The Physicians' Responsibility Toward Hopelessly Ill Patients', *New England Journal of Medicine*, 310(Apr.), 955–9.

Chapter 11

The problem of low benefit/
high cost health care

INTRODUCTION

A widespread perception exists in the United States that we use substantial amounts of high cost/low benefit health care and that this is a major factor in the rapid and seemingly inexorable growth of health care costs. Do our health care dollars often buy life-years of poor quality? Who defines such quality? According to what criteria is it defined? These are the questions I shall address in this essay. The first question is, at least in part, an empirical question which must be settled by relevant data. More specifically, what care is in fact utilized in our health care system, at what cost, and with what effects on patients' lives, are all empirical matters of fact, however limited our data on them may be. Whether the effects of particular health care on patients' lives produce life-years of low quality, and whether that health care represents a poor use of scarce economic resources in comparison with other possible uses of the resources, are both evaluative questions which cannot be settled by empirical data alone. While my expertise lies with the evaluative issues, I want to say something first about the empirical issues.

The term "costworthy care" has been coined by Paul Menzel to refer to health care that is worth its true costs to the patient who receives it.[1] This notion more accurately and broadly captures the issue raised in the first question of whether our health care dollars are buying life-years of low quality. If we are buying life-years of low quality, but of sufficient quality that patients both want the health care necessary to sustain those life-years and find its true costs sufficiently low to make it well worth its benefits, then it is care whose availability and funding should not be limited. So our first question should be understood as a shorthand formulation of the question whether our health care dollars are buying care that is not costworthy.

Even this formulation is still oversimplified in one respect that needs to be noted at the outset. Relatively little health care might be worth its true costs to persons in conditions of extreme economic deprivation because virtually all their resources must be expended on other basic survival needs such as food and shelter. The determination of whether particular health care is costworthy should be made in conditions where persons have a fair or just income share when comparing that health care to other goods. An example of a very crude approximation of that question is: are our health care dollars buying substantial amounts of health care that middle-class Americans would judge to be noncostworthy?

The widespread belief that the answer to this question is affirmative undergirds the common perception that there is a cost containment crisis in health care that will require rationing of health care. (I shall largely avoid the notion of "rationing" in this essay because to many people it connotes denying goods or services to people willing and able to pay for them. Instead, I shall usually speak of limiting the availability or funding of noncostworthy care, leaving open how this might be carried out.) Some who believe there is a cost containment crisis simply point to the total sums of money spent in government programs like Medicare and Medicaid in times of large governmental budget deficits and fiscal constraints. Others point to the rapid rate of growth of health care expenditures beyond the rate of inflation, for example, as a proportion of total GNP or as reflected in increases in health insurance rates paid by employers. Yet, as Henry Aaron and William Schwartz have pointed out, the proportion of GNP absorbed by computers may have increased at least as rapidly as health care during recent decades, but there is no general perception of a cost control crisis in computers.[2] The difference is that in health care it is widely believed that we are utilizing much health care that is not costworthy, that is, whose true costs are too high to justify its benefits. This use of health care is wasteful, not in the sense that the care promises *no* benefits to patients, but in the sense that it promises less benefit for its cost than other health care, or other non-health care goods and services, would produce instead.

INCENTIVES OF PHYSICIANS AND PATIENTS FOR HEALTH CARE UTILIZATION

Before looking briefly at some of the data that bear on whether this belief is correct, it will be helpful to consider the usual circumstances in which decisions are made to utilize health care. By looking at the incentives to which the participants in health care utilization decisions are subject, we can tentatively determine a priori whether it is reasonable to expect that substantial amounts of noncostworthy care will be utilized, and so whether the belief in undue waste in health care is likely to be found correct.

No comparable waste and use of noncostworthy goods are believed present in consumer purchases of other goods such as automobiles, so why expect them in health care? With autos, consumers are generally able to gather information about the features of alternative models, to weigh how well those features fit their particular needs and desires given the costs of alternative models and of other goods and services on which they might instead spend their money. And car buyers, by necessity, do consider the car's full costs in comparison with its benefits because they bear the full costs out of pocket when they buy, or as they pay off their car loan.

Unlike car buyers, patients are commonly believed to be in a relatively poor position to act as informed and prudent purchasers of health care. They usually lack the knowledge, training, and experience to evaluate what health care, if any, they need, whether it will be rendered at reasonable cost, what alternative forms of care are possible, with their attendant benefits, risks, and costs, and even whether the care they receive has been competently rendered. When seriously ill, patients are also commonly fearful, anxious, and dependent, thereby impairing even their usual abilities to decide about health care. The result is that patients often have strong desires to place their care in the hands of physicians whom they can trust to care for them, and so physicians in turn commonly have substantial effective control over many decisions to utilize health care.

What then are the main economic, professional, and ethical incentives of physicians when decisions are made to utilize health care? Though economic systems of reimbursement to physicians have been undergoing substantial change in recent years, most physicians in the United States remain in some form of fee-for-service practice. They earn their living from selling health care, are directly reimbursed for each service performed, and so have an economic incentive to have their patients use more health care, including care which may not be costworthy. Perhaps the most striking evidence of the effects of this economic incentive, though other factors may have been operative as well, is a recent study which found that physicians who did diagnostic imaging for their patients in their own offices and so profited from the service obtained imaging examinations 4 to 4.5 times more often than physicians who always referred their patients to radiologists for imaging examinations and so did not profit from the service.[3] This economic incentive combines with a common understanding of physicians' professional and ethical responsibility to their patients, as expressed in commonly accepted normative accounts or ideals of the physician/patient relationship. According to this patient-centered ethic, physicians' first and foremost responsibility is to their patients – physicians must set aside their own and others' interests in order to do everything possible of potential net benefit to their patients. A crucial part of this patient-centered ethic, as it is understood by many physicians, is that physicians must do everything of potential benefit for their patients *without regard to costs*. The result is that economic and professional or ethical incentives on physicians commonly converge to cause them to render noncostworthy care, including in the extreme case all care of possible positive medical benefit to the patient regardless of its cost.

Other factors, such as fear of malpractice litigation and discomfort with uncertainty, reinforce these incentives for overutilization. It might be objected that many physicians now practice in prospective-payment, capitation-funding settings and managed care programs which reverse the economic incentives of physicians – incentives are now to provide less, not more, care to patients. However, in part because the professional standard of care remains substantially set by the patient-centered ethic which largely disregards costs, capitation-funding schemes such as DRGs, HMOs, and managed care programs have been disappointing to those who expected large cost savings from them. Finally, with payment for health care services coming commonly from insurers, either private or governmental, the economic incentive of physicians and other health care providers for overutilization is reinforced by the assurance that they will be paid for services rendered, even if the patient might be unable to pay for them out-of-pocket.

Patients are the other participants besides physicians in typical decisions to utilize health care. They also typically have a comparable incentive to utilize noncostworthy care. The great majority of patients have some form of health insurance which requires them to bear only limited out-of-pocket costs for their health care. In the extreme, but increasingly rare, case of health insurance with full first-dollar coverage and no co-payments or deductibles, the patient has no out-of-pocket costs. The only economic incentive for such patients not to use all health care of any positive net expected benefit to them is the usually negligible effect on their future health insurance costs, which are usually paid largely or wholly by their employers in any case. In the more common case in which patients' health insurance requires some co-payments and deductibles, they have some limited economic incentive to consider the costs of their health care, but only by the usually small fraction of total actual costs which represents their out-of-pocket copayments and deductibles. And, of course, apart from the economic costs to patients of their health care, it is fully rational for patients to want any and all health care that has any expected net positive benefit to them when benefits are weighed against possible harms of the care, but not also against its costs.

The upshot is that both physicians and patients, deciding together about health care for the patient, have strong economic and other incentives not to consider the true costs of care at all, or at most to give those costs only limited weight, in comparison with the benefits of care. Third party payers and others with incentives to limit the use of health care and its attendant costs are increasingly inserting themselves into health care utilization decisions, but they provide at most limited constraints on physicians' and patients' decisions. Since both physicians and patients have incentives to use substantial amounts of noncostworthy care, the result of overutilization of health care versus other goods and services, and more specifically the use of health care dollars to buy poor-quality life-years, should hardly come as a surprise.

SOME EXAMPLES OF OVERUTILIZATION OR THE USE OF NONCOSTWORTHY CARE

As I noted at the outset of this essay, the most direct answer to the question of whether health care dollars often buy non-costworthy health care will consist in marshalling relevant data concerning health care utilization and its effects on patients' lives. I will not attempt to gather systematically all the available data bearing on this question – that is a very large task for empirical health care researchers and policy analysts, not bioethicists or philosophers. My main concern will be with the two subsequent questions of who should define the benefits, or the quality of life-years produced, and by what

criteria. However, I do want at least to cite some representative evidence that significant health care expenditures do in
fact purchase noncostworthy care, but at the same time to
bring out some of the severe limitations in the data bearing
on this point.

The incentives on the parties to decisions about health care
utilization that I have described above and that can be expected to lead to utilization of noncostworthy care have their
effect virtually across the board of health care, and are not
restricted to care that extends lives of low quality. Nevertheless, there is some reason to believe that the problem of
the use of noncostworthy care in the United States is exacerbated in treatments which produce low quality-adjusted
life-years (QALYs).[4] Cross-cultural data are particularly interesting on this point. Patients with incurable forms of cancer are roughly five times as likely to receive chemotherapy
in the United States as in the United Kingdom.[5] This chemotherapy is commonly highly toxic, with side effects that
usually severely reduce the quality of patients' lives in the
period before their death. Thus, not only is treatment employed whose probability of extending life is very low, but
the treatment itself contributes to substantially reducing the
quality of patients' period of remaining life. This difference
in rates of treatment cannot be laid to overall differences in
health care spending. The forms and rates of treatment of
curable cancers (such as certain cancers of the testis, some
childhood cancers, Hodgkin's disease, and lymphatic leu-
kemia in children) show no appreciable differences across
the Atlantic.

Another well-known difference in treatment between the
United States and the United Kingdom is in the use of dialysis
for chronic kidney failure. Aaron and Schwartz note that
while the rate of patients in the population with functioning
kidney transplants is almost identical (57 people per million
in 1980 in the United States, 56 per million in the United
Kingdom), the rate of patients undergoing dialysis is over
three times as high in the United States (230 versus 69 people
per million).[6] Several points are important about these data.

First, transplantation largely restores the patient's quality of life close to normal – that is, to what it was before kidney failure. Thus, when the quality of life-years gained is high, rates of treatment do not vary, despite the fact that the overall level of health care expenditures per person in the United Kingdom remains substantially below that of the United States. Second, renal dialysis, by comparison, significantly impairs the quality of life of patients, to the extent that one study found that 22 percent of deaths of patients on renal dialysis resulted from the patient's decision to stop the treatment.[7] Thus, renal dialysis is a costly treatment that substantially impairs patients' quality of life, and it is employed at a much higher rate in the United States. Third, it is noteworthy that in the United States dialysis treatment has had financial limitations on access to it virtually eliminated since 1972 by a special funding program set up under the Social Security Administration. I wish to underline that I do not intend any support of the age limit for receiving dialysis which apparently is used in Great Britain – while the United States probably makes significant use of dialysis that is not costworthy, the British deny it in circumstances where it is costworthy.

A third area of life-sustaining treatment in which the data unquestionably support the claim that our dollars are buying life-years of low quality is the treatment of patients in a persistent vegetative state. This is a condition in which the patient suffers from total and permanent unconsciousness, having lost all capacity for conscious experience of any kind. The data are poor on how many such patients exist at any one time, but the President's Commission for the Study of Ethical Problems in Medicine in 1982 estimated the figure for the United States to be about 10,000. If annual costs of care of such patients were to average only $40,000 (I am not aware of data on these costs, but they are often much higher than this), the total annual costs would be about $400 million. The permanent absence of any conscious experience makes it plausible that the quality of life of these patients is not just low but nonexistent; they completely lack any experience of a life. Yet such patients can and often do have their lives maintained in such a condition for many years.

A vastly larger group of persons for whom our health care dollars buy lives of very low quality are patients with severe and permanent cognitive impairment, especially senile dementia of the Alzheimer's type. This disease progressively, often over a number of years, attacks cognitive function and especially memory. Its rate of progression is variable and its effects on patients over time range from only very limited cognitive impairment in early stages of the disease, which allows substantial continued functioning by the patient, to persistent vegetative state in the disease's end stage. While individuals disagree about at what stage in the progression of the disease its effects have become such as to make life no longer worth living, or even considerably worse than death, for most patients there is such a point in the disease. Even before that point, there often is a substantial period of time during which patients' quality of life is such that they would not have chosen before the onset of the disease to devote funds to insure themselves for life-sustaining care instead of using those funds while still healthy for activities and expenditures bringing greater satisfactions.[8]

The provision of intensive care should not go unmentioned even among what are only examples of noncostworthy care. There is probably no other area of modern medicine about which physicians offer more anecdotal reports of care they judge to be highly wasteful in the limited benefits it produces at very substantial cost for some dying patients. My comparisons above of rates of treatment in the United States and Great Britain do not imply that the British have correctly adjusted their provision of care just at the level of costworthy care. There is, of course, no reason to believe that. But the comparisons do illustrate the very large differences in levels of some marginally beneficial care that can result when a society more explicitly confronts limitations in the resources available for health care. Regarding intensive care beds, Aaron and Schwartz estimated that the United States in 1980 had five to ten times as many beds per capita as in the United

Kingdom, and that the British would have had to raise their overall health care expenditures by about 10 percent to reach our level of provision of intensive care. Rationing of intensive care beds has received more explicit attention than rationing in many other areas of health care because hospitals have only a limited number of ICU beds at any one time.[9] Intensivists also have done better at developing measures of potential benefit, such as the APACHE scoring system for admitting patients into ICUs. They have done less well, however, at developing measures and practices for discharging patients whose prognosis has so worsened that they are no longer likely to benefit from ICU care – for these patients, too often, the ICU becomes de facto where they go to die.

Finally, when considering whether significant amounts of noncostworthy health care is provided in the United States, it is a mistake to focus only on examples like those I have given above of treatment that may be relatively directly life-sustaining. Employing again the contrast with the United Kingdom, Aaron and Schwartz estimated that in 1980 the United States had about three times as many CT head scanners and ten times as many body scanners. Even with basic diagnostic x-rays, the United States performs twice as many x-ray examinations per capita and uses twice as much film per examination. It is, of course, much more difficult to measure the effect of this higher use of diagnostic procedures on QALYs produced for patients, and I make no attempt here to determine whether it is costworthy. But the example of diagnostic imaging is an important reminder that incentives for overutilization are pervasive throughout health care and are not limited to more dramatic examples of life-sustaining care.

Having cited examples of costly care that often produces low or no benefit, or more broadly care that is sometimes noncostworthy, several important limitations in this data must be underlined. First, my use of the notion of costworthy care may have given the impression that this standard is both well defined and uncontroversial. Of course, it is neither. Costworthiness is not well defined because, among other

334

reasons, the circumstances in which an individual is to make the judgment that particular care is or is not worth its costs have not been specified – for example, what information the individual has about the effects of the particular care in question, about his or her own health care needs, and what income the individual is assumed to have for the purposes of making this judgment. The standard of costworthiness is also controversial in at least two senses. First, it is controversial how the circumstances for making judgments of costworthiness should be specified in order to make the notion well defined. Second, if these conditions are spelled out to make the notion of costworthiness precisely defined, individuals will still sometimes disagree about whether particular health care employed in particular circumstances would be costworthy. This disagreement reflects different values people place on particular health care benefits in comparison with other health care and non-health care goods and services.

A second limitation in data about the use of noncostworthy care concerns our knowledge of the effects of employing particular forms of health care for particular kinds of patients. It is widely accepted that for most forms of health care these data are severely limited. The unwillingness of many physicians to share fully with their patients the degree of uncertainty about treatment results in turn in a failure of the public to appreciate fully this limitation in medical knowledge.[10] However, the very substantial recent interest in so-called "effectiveness" and "appropriateness" research, matched by a commitment in federal agencies to increase financial support for this research, provides hope for beginning to improve the relevant data and to reduce this limitation.

The third limitation in knowledge about costworthiness is that even with substantial new effectiveness data, the complexity of medicine and of persons will leave considerable residual uncertainty about the effects on any one individual of specific health care interventions in specific circumstances. In advance of actually employing a particular treatment, our knowledge of its effects on a particular individual will inevitably remain probabilistic, providing statistical frequencies for the various effects of alternative treatments for a particular medical condition, in comparison with leaving it untreated. Different evaluations of possible outcomes, as well as different attitudes toward risk and toward risking specific outcomes, will yield different judgments by different individuals about whether particular health care is costworthy. This disagreement emerges clearly in the case of treatments with a low probability of being life-sustaining, such as the use of experimental chemotherapy for a generally incurable form of cancer.

It is difficult to overestimate the formidable difficulties these limitations pose to making and applying judgments about the costworthiness of health care. Nevertheless, I believe the general conclusion stands that our health care dollars buy substantial amounts of noncostworthy care. This then raises the other two questions to which this essay is addressed: Who defines quality of life? By what criteria?

WHO SHOULD DEFINE QUALITY OF LIFE AND COSTWORTHINESS?

At the outset it is important to emphasize that who should define quality of life depends on the use to which the quality of life judgment will be put. The same is true for judgments of costworthiness. By far the most common context in which quality of life judgments are made within health care occurs when treatment decisions about the care of individual patients require evaluation of alternative treatments (including the alternative of no treatment) for the purpose of deciding which alternative promises the best balance of expected benefits over risks for the patient. In most such decisions, the costs of the alternative treatments are not a significant factor because they will be borne by third party payers, commonly private or public health insurance programs.

In recent years a consensus has emerged that this decision should be made in a process of shared decision making between physician and patient. This recognizes the unique

contribution each will normally make to the process of determining which treatment will be best for this particular
patient.[11] The doctrine of informed consent, accepted in both
the law and medical ethics as a requirement for proceeding
with treatment, implies as well that when there is unresolved
disagreement about treatment the competent patient retains
the right to refuse any treatment and to select among available alternative treatments. Thus, the competent patient's
evaluation of alternative treatments and their likely effects
on his or her life must ultimately be respected, though this
is not to say that the patient's evaluation of which alternative
will be best cannot be mistaken.[12] Lodging decisional authority ultimately with the patient is usually thought to be
based on two central values: patient well-being and patient
self-determination. It rests on patient well-being because
competent patients who have been well informed by their
physicians are usually the best judges of what will best serve
their overall good or well-being and the aims and values that
guide their life. It rests also on patient self-determination
when this is understood as people's interest in making important decisions about their lives for themselves and according to their own values.

Who should determine quality of life judgments about alternative treatments does not change when the costs are not
fully borne by third party payers and patients must bear some
out-of-pocket costs, for example through copayments or deductibles in their health insurance. The only difference then
is the additional effect on patients' lives of different dollar
costs for different treatments, and so different alternative
expenditures that must be forgone. The same values of patient well-being and self-determination support shared decision making here, with patients retaining the ultimate right
to refuse any treatment, and to select any available alternative
treatment whose out-of-pocket costs they can pay.

This model of shared decision making, whether for competent patients or as extended to surrogates deciding for
incompetent patients, should not be understood by itself to
create entitlements to the resources necessary to pay for the

337

treatment chosen.[13] What level of health care should be available to persons as a matter of justice, and funded when necessary by public monies, must be established independently by appeal to a general theory of justice, a specific account of justice in health care, and its detailed specification by just political institutions and processes.[14] I shall follow recent discussions in calling this an "adequate level" of health care that should be secured for all members of society.[15] The fundamental question then is when particular health care secures a quality of life for the patient sufficiently low in comparison with its costs that it is not a part of this adequate level. Such health care could be justly distributed on the basis of patients' ability to pay for it.

Who should make judgments about the quality-adjusted life-years secured by particular health care for the purposes of establishing whether patients will have entitlements to that care depends also on the structure of the health care system in which those judgments will be made and applied. In most countries which have national health insurance (NHI) or a national health service (NHS) providing the great bulk of health care used in the country, it is possible to set overall budgets, first at a national level and then at increasingly local levels. Health planners at local levels then can make tradeoffs at the margins regarding alternative expenditures, including not only provision of care but also capital investments in facilities and technology. Various degrees of public involvement in decisions made at these different levels are also possible.

An NHS or NHI has one extremely important political, and in turn ethical, advantage for defining and funding cost-worthy care. When most of the population secures health care through either NHI or an NHS, the coverage must be sufficiently extensive to be viewed as adequate by most of the population. If it is not, political pressure to expand that coverage, together with increased purchase of health care outside the NHI or NHS program, can be expected. Inclusion of most citizens under NHI or an NHS can be crucial to keeping the level of benefit coverage at an adequate level.

Public programs serving only disadvantaged minorities are much more vulnerable to cost containment and budget pressures resulting in denial of costworthy care, since the denial does not affect most members of the society. This was one of the most serious initial ethical concerns about the much publicized, recently proposed changes to limit treatment coverage in the Oregon Medicaid program, though it has subsequently become clear that this benefit package will apply well beyond members of the Medicaid program.

In the United States, for at least the next several years, the size of projected federal budget deficits together with other political opposition will make NHI or an NHS difficult to achieve.[16] So long as we retain our extremely heterogeneous organizational and funding system in health care, it will inevitably require different persons and procedures, located in different parts of the overall health care system, to define noncostworthy care. Because the great bulk of that care will be financed through some form of public or private insurance program, it is not possible to use a market to fully individualize decisions about what care is costworthy. Moreover, even in the absence of insurance, health care markets would be poor procedures for determining costworthy care. I believe it is possible, however, to identify several desirable features of ethically defensible procedures for identifying noncostworthy care that need not be funded for patients: (1) decisions and guidelines for identifying such care should be arrived at through public processes that allow substantial input and participation to those who will be affected by the decisions; (2) procedures for implementing these decisions and guidelines must be sufficiently flexible to reflect the widely variable benefits for different patients in different conditions that a particular kind of service or procedure can have – this will inevitably require a role for physicians in exercising their professional judgment in the application of guidelines to their patients;[17] (3) health care institutions limiting the availability or funding of noncostworthy care should fully inform current and prospective patients of those limitations; (4) procedures must be developed to monitor over time the application of limitations on the availability or funding of noncostworthy care to ensure that it is done equitably and without denying patients access at least to an adequate level of care.

The appropriate persons and decision-making bodies for defining limitations on the provision or funding of noncostworthy care will usually vary depending on the context. For example, in an HMO guidelines excluding provision of non-costworthy care might be addressed by a committee within the HMO with substantial patient member representation. Such representation is especially important since both HMO administrators as well as employers who pay for the members' insurance have financial interests in conflict with members' interests, and so have financial incentives to overestimate the cost savings and undervalue the lost benefits to patients from excluding marginal care from coverage. For government insurance programs, open public debate at appropriate points in the political and policy process, such as in legislatures, public hearings, and so forth, would be appropriate. In other cases, participant input can be fostered by employers and health insurers providing their employees and insurees with a greater range of alternative insurance plans that attempt to define and limit noncostworthy care to varying extents. This will require major new efforts in the very difficult educational task of making the nature and impact of the different alternative choices clear and understandable to employees. Indeed, if the public is to have any realistic hope of effectively representing its interests in any of these processes, major new public educational efforts are needed on the nature of the alternatives and tradeoffs. The general point is that in a highly heterogeneous health care system such as our own, no particular persons or single institutional mechanism can claim sole legitimacy to address and make decisions about what care will not be provided or funded to patients because it is not costworthy.

Nevertheless, there are some institutional mechanisms for defining costworthiness that need not vary depending on the context in which the guidelines for costworthiness are to be applied, but might be legitimately employed in nearly all decision-making contexts about treatment utilization. For example, consensus conferences have been convened by a variety of groups in several countries in recent years to establish treatment guidelines for particular conditions.[18] Consensus conferences have usually not given a significant, or sometimes even any, role to cost in developing recommendations regarding treatment alternatives, and have not focused specifically on determinations of costworthy care. But if their membership were broadened appropriately to include representatives of patients and the public, such conferences could explicitly expand their concern to address costworthiness, or at least the different but not unrelated issue of cost effectiveness of treatment alternatives. Likewise, what has been described as the "new technology assessment," which goes beyond the usual restriction to a biomedical perspective to add social, economic, and ethical perspectives, could explicitly address costworthiness.[19]

WHAT CRITERIA SHOULD BE USED TO DEFINE QUALITY OF LIFE AND COSTWORTHINESS?

The appropriate persons and procedures for making and implementing decisions about quality of life and costworthiness must inevitably vary because of the great institutional heterogeneity of our health care system. Nevertheless, there might be objectively correct criteria by which all should make these decisions. However, I believe there is no single set of objectively correct and relatively precise criteria for defining the costworthiness of health care, and there is certainly no general agreement about those criteria. These criteria are not just a matter of weighing benefits against costs of health care, but also require full theories of distributive justice and justice in health care. There is no more agreement among experts on these theories than there is among the general public. Determining this entitlement level in a democratic society is appropriately a matter for public debate and choice, not expert determination. Nevertheless, that debate and choice

341

should be informed by the best thinking on distributive justice generally, on justice in health care in particular, and on evaluating the benefits of health care. Evaluating alternative accounts of distributive justice and justice in health care for defining an adequate level of care is beyond the scope of this essay. But criteria for evaluating the benefits of particular forms of health care, and more specifically the effects of health care on the quality of life of patients, can be briefly discussed.[20] Moreover, there is a large and growing literature within health policy analysis addressing the assessment of the effects of health care on the quality of patients' lives.

Early attempts to measure the effects of health and disease on a population were relatively crude measures of morbidity and mortality, as evidenced by such data as life expectancy, infant mortality, and the incidence of various diseases in a population. In recent decades, health policy researchers have developed a variety of measures more sensitive to the effects of disease and its treatment on people's quality of life. As generally conceived, the appropriate aims of medicine are to prevent, palliate, or cure disease and its associated effects of suffering and disability, and thereby to restore or to prevent the loss of normal function or life. Quality of life measures in medicine and health care generally focus on dysfunction and its relation to normal function. At a deep level, medicine views bodily parts and organs, individual human bodies, and persons from a functional perspective. Some measures of the effects of disease and its treatment on quality of life are disease-specific, but for policy purposes in evaluating alternative uses of resources or costworthiness of care these have the great disadvantage that they do not allow comparisons across treatments of different diseases.[21] Other measures have been developed to measure overall health status or quality of life as it has been affected by any disease, and it will be helpful to have two examples before us.

The Sickness Impact Profile (SIP) was developed by Marilyn Bergner and colleagues to measure the impact of a wide variety of forms of ill health on the quality of persons' lives (see Table 10.1, in the preceding chapter, pp. 299–300).[22]

A different measure is the Quality of Well-being Scale (QWB) which "combines preference-weighted measures of symptoms and functioning to provide a numerical point-in-time expression of well-being that ranges from zero (0) for death to 1.0 for asymptomatic optimum functioning"[23] (Table 11.1). In what the authors call the General Health Policy Model, "the QWB inputs are integrated with terms for the number of people affected and the duration of time affected to produce the output measure, which is known as the 'well year'."[24] Dollar cost per well year produced can then be calculated for different alternative health and treatment programs for the purposes of setting policy priorities (Table 11.2).

I will not address here the various methodological issues about how successful such measures are in what they purport to measure.[25] Rather, I will use the examples of the SIP and QWB measures to develop some of the ethical issues in their use as possible answers to one of the questions of this essay: What criteria should be used to measure the benefits, that is, the effects on quality of life, of health care expenditures? First, it is important to underline that even an SIP score or a dollar cost per well year for a health care program or form of treatment for a particular disease will not tell us whether that program or treatment is costworthy or should be part of an adequate level of care. For that, we need the threshold cost above which either SIP improvements or well years are no longer costworthy. No general theory of justice, or theory of justice in health care, provides anything even approaching a precise answer to what these thresholds should be for inclusion in an adequate level of health care that is to be guaranteed for all citizens.

I noted earlier a very rough criterion for determining whether particular marginal treatments or services are costworthy: Would most middle-class persons accept as adequate a health insurance package that excluded coverage of the treatment or service, or would they instead buy supplemental insurance for it or pay for it out-of-pocket?[26] Of course, most people are unlikely to be able to select a precise threshold,

Table 11.1. *List of Quality of Well-being General Health Policy*
Model Symptom/Problem Complexes (CPX)
with Calculating Weights

CPX No.	CPX Description	Weights
1	Death (not on respondent's card)	−0.727
2	Loss of consciousness such as seizure (fits), fainting, or coma ("out cold" or "knocked out")	−0.407
3	Burn over large areas of face, body, arms, or legs	−0.387
4	Pain, bleeding, itching, or discharge (drainage) from sexual organs – does not include normal menstrual bleeding	−0.349
5	Trouble learning, remembering, or thinking clearly	−0.340
6	Any combination of one or more hands, feet, arms, or legs either missing, deformed (crooked), paralyzed (unable to move), or broken – includes wearing artificial limbs or braces	−0.333
7	Pain, stiffness, weakness, numbness, or other discomfort in chest, stomach (including hernia or rupture), side, neck, back, hips, or any joints or hands, feet, arms, or legs	−0.299
8	Pain, burning, bleeding, itching, or other difficulty with rectum, bowel movements, or urination (passing water)	−0.292
9	Sick or upset stomach, vomiting, or loose bowel movement, with or without fever, chills, or aching all over	−0.290
10	General tiredness, weakness, or weight loss	−0.259
11	Cough, wheezing, or shortness of breath with or without fever, chills, or aching all over	−0.257
12	Spells of feeling upset, being depressed, or crying	−0.257
13	Headache, dizziness, ringing in ears, or spells of feeling hot, nervous, or shaky	−0.244
14	Burning or itching rash on large areas of face, body, arms, or legs	−0.240
15	Trouble talking such as lisp, stuttering, hoarseness, or being unable to speak	−0.227
16	Pain or discomfort in one or both eyes (such as burning or itching) or any trouble seeing after correction	−0.230
17	Overweight for age and height or skin defect of face, body, arms, or legs such as scars, pimples, warts, bruises, or changes in color	−0.188
18	Pain in ear, tooth, jaw, throat, lips, tongue; several	−0.170

Table 11.1. (continued)

CPX No.	CPX Description	Weights
	missing or crooked permanent teeth – includes wearing bridges or false teeth; stuffy, runny nose; or any trouble hearing – includes wearing a hearing aid	
19	Taking medication or staying on a prescribed diet for health reasons	− 0.144
20	Wore eyeglasses or contact lenses	− 0.101
21	Breathing smog or unpleasant air	− 0.101
22	No symptoms or problems (not on respondent's card)	− 0.000
23	Standard symptom/problem	− 0.257

Source: Robert M. Kaplan et al., "The Quality of Well-Being Scale: Applications in AIDS, Cystic Fibrosis, and Arthritis," *Medical Care* 27, 3 Suppl. (1989) S27–S43.

or even a relatively limited threshold range, of dollar cost per well year or SIP improvement above which care is not cost-worthy. Based on well-year costs alone, only determinations including relatively inexpensive programs as costworthy and excluding very expensive programs as noncostworthy will usually be possible. For most people, more meaningful judgments would require at least the following: the patient condition(s) in which a treatment is employed, together with the respects in which that condition lowers quality of life; specification of the range and frequency of different effects of a treatment on the quality of patients' lives; the treatment's average cost; the frequency in the population of need for the treatment; the consequent cost of the insurance supplement that covers reimbursement for the treatment. Nevertheless, suppose the QWB is a valid measure of quality of well-being that accurately reflects the relative importance people give to various factors that affect their well-being. Then, once a threshold point has been determined for costworthiness or

Table 11.2. *Quality of Well-being General Health Policy Model and Sample Calculation*

Step No.	Step Definition	Weight
Mobility scale (MOB)		
5	No limitations for health reasons	−0.000
4	Did not drive a car, health related: did not ride in a car as usual for age (15 yr) (health related), *and/or* did not use public transportation (health related), *or* had or would have used more help than usual for age to use public transportation (health related)	−0.062
2	In hospital, health related	−0.090
Physical activity scale (PAC)		
4	No limitations for health reasons	−0.000
3	In wheelchair, moved or controlled movement of wheelchair without help from someone else, *or* had trouble or did not try to lift, stoop, bend over, or use stairs or inclines (health related) *and/or* limped, used a cane, crutches, or walker (health related), *and/or* had any other physical limitation in walking, or did not try to walk as far or as fast as others the same age are able (health related)	−0.060
1	In wheelchair, did not move or control the movement of wheelchair without help from someone else, *or* in bed, chair, or couch for most or all of the day (health related)	−0.077
Social activity scale (SAC)		
5	No limitations for health reasons	−0.000
4	Limited in other (e.g., recreational) role activity (health related)	−0.061
3	Limited in major (primary) role activity (health related)	−0.061
2	Performed no major role activity (health related) but did perform self-care activities	−0.061

Table 11.2. (*continued*)

Step No.	Step Definition	Weight
1	Performed no major role (health related) *and* did not perform or had more trouble than usual in performance of one or more self-care activities (health related)	−0.106

Calculating formulas

Formula 1: Point-in-time well-being score for an individual (W):

$$W = 1 + (CPXwt) + (MOBwt) + PACwt + SACwt,$$

where wt is the preference-weighted measure for each factor and CPX is the symptom/problem complex. For example, the W score for a person with the following description profile may be calculated for one day as follows:

Quality of Well-being Element	Step Definition	Weight
CPX-11	Cough, wheezing, or shortness of breath, with or without fever, chills, or aching all over	−0.257
MOB-5	No limitations	−0.000
PAC-1	In bed, chair, or couch for most or all of day (health related)	−0.077
SAC-2	Performed no major role activity (health related) but did perform self-care	−0.061

$$W = 1 + -0.257 + -0.000 + -0.007 + -0.061 = 0.605$$

Formula 2: well years (WY) as an output measure:

$$WY = [\text{No. of persons} \times (CPXwt + MOBwt + PACwt + SACwt)] \times \text{time}$$

Source: Robert M. Kaplan et al., "The Quality of Well-Being Scale: Applications in AIDS, Cystic Fibrosis, and Arthritis," *Medical Care* 27, 3 Suppl. (1989) S27–S43.

the adequate level, the QWB scale and the dollar cost per well year for different treatments of specific conditions can directly determine whether any particular treatment of a particular condition is costworthy or should be part of the adequate level.

What do these measures tell us about criteria for the quality of life as it is affected by health care, and about some of the ethical issues their use to limit available or funded care would raise? One central issue is the extent to which general measures based on community-wide generalizations about the importance of specific effects on the quality of life of individuals overlook a substantial range of variation either in the impact on the quality of individual lives or in the relative importance different individuals give to these impacts. To what extent should ethically defensible limits on available or funded care be sensitive to these variations? Second, note that the principal emphasis in these measures of the quality of life is on function, and functions of the "whole person" as opposed to body parts or individual organ systems. In QWB this functional focus is in the effect of symptom/problem complexes on mobility and physical and social activity. Functions are broadly characterized so as to be relevant to virtually any life plan common in modern societies; following Rawls's notion of "primary goods," these can be called primary functions. To illustrate just with the SIP, the "independent categories" of sleep and rest and of eating are necessary for biologic function. The categories of work, home management, and recreation and pastimes are central activities in virtually all lives; their relative importance in a particular life is partially adjusted for by making the measure relative to what had been the individual's normal level of activity in each of these areas prior to sickness. The two broad groups of function, physical and psychosocial, are each broken down into several distinct components. For each primary function, the SIP measures the impact of sickness on it by eliciting information concerning whether activities typical in the exercise of that function continue to be performed or have become limited. However, even for primary functions which are plausibly thought to be important in virtually any life, the SIP does not measure the different relative value or importance they have within different lives.

A third important component of any defensible measure of quality of life is individuals' subjective response to their objective physical condition and level of function, or what could be characterized as their level of happiness or satisfaction with their lives. This is captured to a quite limited degree in the emotional behavior category of SIP and in the symptom/problem complex 12 of QWB (spells of feeling upset, being depressed, or crying). The important point is that how good a life a person has, and how significant or valuable a change in that life is, depends in part on how happy or pleased the person is with how the life is going, or with the change in the life. This subjective response or happiness component of quality of life is a response to, and so is not unrelated to, the objective conditions and other functional components of the quality of life. Nevertheless, medicine provides many examples that show it is a mistake to assume this subjective response component correlates closely and invariably with other objective functional measures. For example, two studies of patients with end-stage renal disease found a substantial relation between different objective function variables and also between different subjective response variables, but only a very limited relation between objective and subjective variables.[27] These data illustrate the importance of including both objective function measures and subjective response measures in any comprehensive measure of the quality of life since neither is a reliable surrogate for the other.

A serious problem for quality of life measures to be used across broad populations is the relative weight to be given objective function versus subjective response criteria. Individuals differ considerably on this. In the face of seriously debilitating injury or disease, one patient may adjust her aspirations and expectations to her newly limited functional capacities and place great value on finding happiness and satisfaction in them despite her limitations. Faced with similar debilitation, another patient may assign little value to adjusting to her disabilities, instead not wanting to become the kind of person who adjusts to and is happy in that debilitated and dependent state.

People's adjustments to severe, adult-onset disability which requires substantial changes in central activities like work and social and recreational activities bring out another important component of criteria for evaluating quality of life. When these disabilities significantly restrict the activities that would otherwise have been available to and pursued by people, they will, all other things equal, constitute reductions in their quality of life. But this will be because the disabilities restrict the choices, or what Norman Daniels has called the normal opportunity range, available to people, and not because the compensating paths chosen need be, once fully entered on, any less desirable or satisfying.[28] The opportunity for choice from among a reasonable array of life plans is one important criterion of quality of life. It reflects the importance of individual self-determination, and helps explain the importance of functional capacities that permit choice from among a significant range of alternatives for quality of life.

Opportunities for choice from among reasonable arrays of alternatives, or having available a range of opportunity normal for one's society, are additional important objective components in quality of life evaluations which a focus only on subjective response, satisfaction, or happiness components can miss. An individual's happiness is to a significant extent a function of the degree to which his or her major aims are being at least reasonably successfully pursued. After serious debilitating illness people often come to be satisfied with much less in the way of hopes and accomplishments. They are satisfied with getting much less from life because they have come to expect much less, but it remains true that they do get less or have a less good life.

Limitations on the availability or funding of life-sustaining treatment or expensive treatment and rehabilitative programs for seriously debilitated patients might seem unjustified so long as patients still consider, or will come to consider, the severely constrained life worth living. That would be correct if we are seeking the threshold point at which patients reject life-sustaining treatment because they no longer find the life that treatment sustains worth living, *independent of the costs of the treatment.* But the issue here is whether a treatment has sufficiently limited impact on the quality of patients' lives relative to its costs that it is not reasonably considered costworthy or part of an adequate level of health care – for example, because most middle-income persons would not wish to purchase insurance to cover the cost of that care. Some health care, including life-sustaining care, that people would welcome when it has no cost to them would not be insured for by them when the insurance is priced to the true costs of providing the care.

CONCLUSION

I have sought to address three questions in this paper: (1) Do our health care dollars often purchase life-years of low quality? (2) Who should define that quality? (3) By what criteria should that quality be defined? Both the incentives of the parties to health care utilization decisions, as well as the relevant data, suggest the answer to the first question is yes. It need be neither irrational nor unethical to attempt to reduce the use of noncostworthy care, though, of course, some ways of doing so might be irrational or unethical. The second question permits no single answer. In a highly heterogeneous health care system like that in the United States, the attempt to define and limit use of noncostworthy care will inevitably occur at many places in that system. I have suggested, however, some features of ethically defensible procedures to do this. In response to the third question, I have offered examples of the kinds of criteria that have been developed for evaluating the effects of health care on patients' quality of life, and noted a few of the ethical issues raised by their use in doing so.

Changes in the health care system to reduce the use of noncostworthy care are desirable and should be welcomed.

However, the same powerful social, political, and economic pressures which make further limits likely on the availability or funding of health care in order to control health care costs also give reason for concern that these limits be ethically defensible and not worsen access to an adequate level of care. Indeed, perhaps the strongest defense of such limits would be if they played an essential role in reducing or eliminating the shameful inequities that persist in the United States today in access to even the barest minimum of health care.

NOTES

1 Paul Menzel, *Medical Costs, Moral Choices: A Philosophy of Health Care Economics in America* (New Haven: Yale University Press, 1983).

2 Henry J. Aaron and William B. Schwartz, *The Painful Prescription: Rationing Hospital Care* (Washington, DC: The Brookings Institution, 1984), p. 3.

3 Bruce V. Hillman et al., "Frequency and Costs of Diagnostic Imaging in Office Practice – A Comparison of Self-Referring and Radiologist-Referring Physicians," *New England Journal of Medicine* 323 (1990) 1604–1608.

4 Among recent critical discussions of the use of QALYs in health care program evaluation and policy are John LaPuma and Edward F. Lawlor, "Quality-Adjusted Life-Years: Ethical Implications for Physicians and Policymakers," *JAMA* 263 (1990) 2917–2921; Ichiro Kawachi, Peter Bethwaite, and Judy Bethwaite, "The Use of Quality-Adjusted Life Years (QALYs) in the Economic Appraisal of Health Care," *New Zealand Medical Journal* 103 (1990) 46–48; John Rawls, "Castigating QALYs," *Journal of Medical Ethics* 15 (1989) 143–147.

5 Basil A. Stoll, "What is Overtreatment in Cancer?" in *Cost Versus Benefit in Cancer Care*, ed. B. A. Stoll (Baltimore, MD: Johns Hopkins University Press, 1988); G. J. C. Rees, "Cost-Effectiveness in Oncology," *Lancet* 2 (1985) 1405–1407; Aaron and Schwartz, *op. cit.*, 44–50.

6 Aaron and Schwartz, *op. cit.*, 29–37.

7 Steven Neu and Carl M. Kjellstrand, "Stopping Long-Term Dialysis: An Empirical Study of Withdrawal of Life-Supporting Treatment," *New England Journal of Medicine* 314 (1986) 14–20.

8   A prudential lifespan approach to allocating resources to health care at different stages of one's life is developed in Norman Daniels, *Am I My Parents' Keeper?* (Oxford and New York: Oxford University Press, 1988). I develop the implications of a prudential lifespan approach for funding health care for the severely demented in "Justice and the Severely Demented Elderly," Chapter 12 of this volume.

9   Michael J. Strauss et al., "Rationing of Intensive Care Unit Services: An Everyday Occurrence," *JAMA* 255 (1986) 1143–1146; H. Tristram Engelhardt, Jr., and Michael A. Rie, "Intensive Care Units, Scarce Resources, and Conflicting Principles of Justice," *JAMA* 255 (1986) 1159–1164; Daniel E. Singer et al., "Rationing Intensive Care – Physician Responses to a Resource Shortage," *New England Journal of Medicine* 309 (1983) 1155–60.

10 Jay Katz, "Why Doctors Don't Disclose Uncertainty," *Hastings Center Report* 14, 1 (February 1984) 35–44.

11 Cf. Report of the President's Commission for the Study of Ethical Problems in Medicine and Biomedical and Behavioral Research, *Making Health Care Decisions* (Washington, DC: U.S. Government Printing Office, 1982). I discuss some complexities in the roles of physicians and patients in this process in "The Ideal of Shared Decision Making between Physicians and Patients," Chapter 2 of this volume.

12 I discuss some ways in which patients can be mistaken in Chapter 2 of this volume and in Dan W. Brock and Steven A. Wartman, "When Competent Patients Make Irrational Choices" (Chapter 3 of this volume), where physicians' responsibilities in the face of bad choices by their patients are also explored.

13 Cf. Allen E. Buchanan and Dan W. Brock, *Deciding For Others: The Ethics of Surrogate Decision Making* (Cambridge: Cambridge University Press, 1989), ch. 4.

14 The most extensive and best treatment of justice in health care is Norman Daniels, *Just Health Care* (Cambridge: Cambridge University Press, 1985).

15 Report of the President's Commission for the Study of Ethical Problems in Medicine and Biomedical and Behavioral Research, *Securing Access to Health Care* (Washington, DC: U.S. Government Printing Office, 1983).

16 In this and the next two paragraphs I draw on my discussion in "When is Patient Care Not Costworthy? The Case of Ger-

trude Handel," in *Casebook on the Termination of Life-Sustaining Treatment and the Care of the Dying*, ed. Cynthia B. Cohen (Bloomington, IN: Indiana University Press, 1988), p. 139.

17 Cf. William B Schwartz and Henry J. Aaron, "The Achilles Heel of Health Care Rationing," *New York Times* July 9, 1990.

18 Cf., for example, Jonathan Lomas et al., "Do Practice Guidelines Guide Practice? The Effect of a Consensus Statement on the Practice of Physicians," *New England Journal of Medicine* 321 (1989) 1306–1311; J. Vang, "The Consensus Development Conference and the European Experience," *International Journal of Technology Assessment in Health Care* 2 (1986) 65–76; J. Kosecoff et al., "Effects of the National Institutes of Health Consensus Development Program on Physician Practice," *JAMA* 258 (1987) 2708–13.

19 Victor R. Fuchs and Alan M. Garber, "The New Technology Assessment," *New England Journal of Medicine* 323 (1990) 673–677.

20 I address a broader array of quality of life measures in much greater detail in "Quality of Life Measures in Health Care and Medical Ethics," Chapter 10 of this volume, on which this section draws extensively.

21 An example of disease-specific measures is R. F. Meenan, E. H. Yelan, and J. H. Mason, "Measuring Health Status in Arthritis: the Arthritis Impact Measurement Scales," *Arthritis Rheumatoid* 23 (1980) 146–156.

22 Marilyn Bergner et al., "The Sickness Impact Profile: Conceptual Formulation and Methodology for the Development of a Health Status Measure," *International Journal of Health Services* 6 (1976) 393–415; and Marilyn Bergner et al., "The Sickness Impact Profile: Development and Final Revision of a Health Status Measure," *Medical Care* 19 (1981) 787–805.

23 Robert M. Kaplan et al., "The Quality of Well-Being Scale: Applications in AIDS, Cystic Fibrosis, and Arthritis," *Medical Care* 27, 3 Suppl. (1989) S27–S43.

24 *Ibid.*, S32. Cf. also Robert M. Kaplan and John P. Anderson, "A General Health Policy Model: Update and Applications," *HSR; Health Services Research* 23 (1988) 203–235.

25 An excellent review of these issues is in a series of four papers by Debra G. Froberg and Robert L. Kane: "Methodology for Measuring Health State Preferences – I: Measurement Strategies," *Journal of Clinical Epidemiology* 42, 4 (1989) 345–354;

"Methodology for Measuring Health State Preferences – II: Scaling Methods," *Journal of Clinical Epidemiology* 42, 5 (1989) 459–471; "Methodology for Measuring Health State Preferences – III: Population and Context Effects," *Journal of Clinical Epidemiology* 42, 6 (1989) 585–592; "Methodology for Measuring Health State Preferences – IV: Progress and a Research Agenda," *Journal of Clinical Epidemiology* 42, 7 (1989) 675–685.

26 Mary Ann Baily, "Rationing Medical Care: Processes for Defining Adequacy," in *The Price of Health*, eds. G. J. Agich and C. E. Begley (Dordrecht, Holland: D. Reidel Publishing Company, 1986).

27 Roger W. Evans, "The Quality of Life of Patients with End Stage Renal Disease," *New England Journal of Medicine* 312 (1985) 553–559; O. Lynn Deniston et al., "Assessment of Quality of Life in End-Stage Renal Disease," *HSR: Health Services Research* 24 (1989) 555–578.

28 Daniels, *Just Health Care, op. cit.*

Chapter 12

Justice and the severely demented elderly

I. INTRODUCTION

This essay addresses a narrowly circumscribed aspect of justice and the elderly. What health care and expenditure of resources on health care are owed on grounds of justice to the severely demented elderly? This is not an entirely accurate specification of the group of patients with whom I am concerned in several respects. In the great majority of cases, the severely demented are among the elderly, understood here as the over age 65 population, but in a minority of cases dementia can progress to this stage in younger persons. In that respect, my argument here will apply to the claims to care of some non-elderly as well. The effects of dementia that are my special concern here are the erosion of memory and other cognitive functions that attack and, I shall argue, ultimately destroy personal identity and personhood in the patient. While senile dementia of the Alzheimer's type is probably the most common cause of cognitive disability of the specific form that erodes personal identity, it may have

For helpful comments on an earlier draft of this essay, I would like to thank Grant Gillett, Stephen L. White, and most especially Allen Buchanan who provided very detailed and useful comments. The issues discussed in this essay, together with related issues, are addressed at greater length in Buchanan and Brock (1986), originally commissioned by the Office of Technology Assessment, U.S. Congress, and in our book *Deciding For Others: The Ethics of Surrogate Decision Making* (Cambridge University Press, New York, 1989).

other causes as well. Thus, the implications of my argument here will extend to some other severely cognitively disabled patients besides the severely demented, though for convenience I shall generally refer simply to the severely demented.

I shall argue that because of the crucial role played by memory and other forms of psychological continuity in maintaining the identity of a person through time, the loss of personal identity occurs before and in the absence of a complete and irreversible loss of consciousness, or what is sometimes called a persistent vegetative state (PVS). However, I believe that since personal identity and personhood are also destroyed in patients with permanent and complete loss of consciousness, an argument that turns on the destruction of personal identity will apply also to these patients. While the complete and irreversible loss of consciousness often occurs in end-stage demented patients, it is probably more common in cases of severe trauma or insult to the brain from accidents and major stroke. Here too, and even more often, my argument will apply to some non-elderly as well as some elderly. The class of patients with which I am concerned is thus, thankfully, only a small proportion of the elderly, and also includes some non-elderly. Nevertheless, the bulk of patients in whom personal identity and personhood are destroyed are elderly, and the size of this group of patients can be expected to increase in coming years in the absence of new means to prevent or cure dementia and with the increasing aging of the population, in particular with the growth of the so-called "old-old."

Quite obviously, I cannot attempt to develop here a full theory of justice and health care, with its implications for justice between generations and for the severely demented and cognitively disabled. Instead, I will seek briefly to show that both of the most common approaches to these questions place special weight on whether personal identity and personhood are maintained in the parties to whom justice is owed. More precisely, we shall see later the necessity, for some questions about just claims to health care, of distinguishing whether personal *identity* is destroyed from whether *personhood* is destroyed.

One quite plausible and increasingly common approach to these questions focuses on the problem of justice between generations or different age groups. It exploits the crucial fact that each person who achieves a normal lifespan passes through all of the different age groups as he or she ages, and then asks of a single individual who is assumed to have a fair lifetime income share, how much of his or her resources that person would allocate to health care, and to different kinds of health care, at different stages of the person's life.[1] Many important details and restrictions on this choice problem need to be spelled out if it is to play a plausible role in a theory of justice and health care. I shall not enter into these specifications here. However, once spelled out I believe this approach does clearly yield the result that persons would allocate different shares of their resources for health care needed at different stages of their lives because of different expected needs and different potential benefits of health care at different life stages. Two important differences often, though of course not always, associated with age are the amount of expected life years that life-sustaining treatment may make possible, and the expected quality of the life years sustained.[2] The quality of a person's life is affected adversely by virtually all levels of dementia and cognitive disability. Also important are differences in characteristic individual needs and desires at different stages in the life cycle that can be affected by health care.

What is central to this approach is that the problem of the allocation of resources among the elderly and other groups is transformed from a distributional problem for different competing individuals into a prudential problem for individuals of how they would allocate their own resources among different stages of their own lives. Because in the usual case each individual can expect to pass through the different stages of life, it is possible to transform what appears at any particular point in time to be a zero-sum problem of distributive justice in allocating resources among different competing persons or groups – the old and the young – into a problem of prudential reasoning for individuals in allocating their own resources to different stages of a single, that is their own, life. This permits avoiding the difficult and controversial questions in interpersonal distributions about the extent to which losses to some persons can be justified by gains to other persons. In the normal case, each of us will pass through the different age groups in the course of a life, unlike the case for different races, genders, and so forth.

One important point for my purposes is that this prudential allocator approach seems clearly to presuppose (but see the qualification in Section IV below) that the personal identity of the allocator is maintained throughout the different stages of the single life to which resources are to be allocated. Only then can the allocation problem be framed in terms of the prudential reasoning of a single person concerned only with outcomes for him- or herself at different stages of his or her life.[3]

The other broad approach to the problem of the just claims to health care of the severly demented elderly also focuses on the moral importance of this group's severe cognitive disability, but accepts the interpersonal nature of the young–old conflict over resources posed by the elderly. This approach retains the assumption that the problem is one of competition at any given point in time among different groups of persons for limited and scarce health care resources. There is little consensus about how the problem should be resolved under what I shall call an interpersonal distribution approach because there is no consensus about general theories of distributive justice, nor about theories of justice in the allocation of health care. All plausible theories, however, must acknowledge that resource scarcity implies individuals' rights to health care, or society's obligations to insure access to health care for its citizens, are limited to something less than all health care of some positive expected benefit without regard to its cost. Such theories then generally focus on properties that make health care of special importance for persons: its role in preventing disability and preserving opportunity, in preventing or relieving suffering, and in extending life.[4] Implicit in virtually all such accounts is that only persons have any rights against others that health care be provided to them, or are the object of our obligations to secure an adequate level of health care. I shall go into the question a bit more below, but I note here simply that it is properties possessed by persons and not shared by other beings that make plausible any claims on the efforts of others to preserve an individual's opportunities or extend his or her life. Obligations or rights concerning the relief of suffering, however, may extend to non-persons and so in turn, as we shall see, may claims regarding health care.

I have contrasted above a prudential allocator approach to the problem of the just claims to health care of patients within different age groups and of different cognitive status, which analyzes the problem in terms of a single individual allocating his or her resources over a lifetime, from an interpersonal distribution approach, which analyzes it as a problem of the distribution of scarce resources to competing groups composed of persons of different ages and cognitive status. However, there is substantially more connection and overlap in these two approaches than this contrast implies. The effects of health care on function and opportunity, relief of suffering, and prolongation of life that constitute its special importance and in the interpersonal distribution approach ground the social obligation to insure access to an adequate level of health care for all, are the same effects that prudential allocators will attend to in deciding how to allocate their resources to health care across their lives. Thus, we can expect that *both* approaches will focus on the effects of health care on preserving function and opportunity, relieving suffering, and prolonging life in determining just claims to health care. As a result, in what follows I will sometimes draw out the implications only of the prudential allocator approach, leaving it to the reader to fill in the closely parallel reasoning of the interpersonal distribution approach.

II. THE JUST CLAIMS TO HEALTH CARE OF THE PATIENT WHO HAS DIED

In order to explore the implications of severe dementia for just claims to health care, I want to begin with the uncontroversial, but I believe instructive, case of the patient who has died. Most states now have two criteria for death: either so-called brain death, in which all brain activity, including that of the brain stem, has ceased, or the irreversible cessation of heart and lung function. Such whole brain criteria for death were introduced in response to inadequacies of the older heart/lung definitions for patients whose heart and lung functions were being artificially maintained by mechanical means but in whom all brain activity had irreversibly ceased. Such patients are properly determined to be dead.[5]

What claims do the dead have to continued health care? It is widely agreed that there is *no* obligation to use *any* social resources to maintain bodily functions in a person who is now dead, at least for the sake of that person. In fact it is widely acknowledged that with death the person has ceased to exist and that only the body of the person who once existed now remains. Respectful treatment and disposition of the body for the sake of those who cared for the person who had died, and out of respect for the wishes of the patient before his or her death, are expected. However, once this person has died, it is no longer possible to use health care for any of the purposes it serves for persons. Life is gone and with it all possibility for extending life, preserving or restoring opportunity, or relieving suffering; nothing that health care can now do for the body makes possible any benefit to the person who has ceased to exist.

If a person while alive and competent asks that all efforts be made to artificially maintain his bodily functions after his death, our society refuses to use social resources to do so, and health care professionals refuse to employ their efforts to do so as well. None of the functions or goals of health care of treating disease in order to promote opportunity, relieve suffering, or preserve life can be furthered once the person has died.[6]

Physicians do not consider that there is any health care decision-making left to a patient's surrogate decision-makers once the patient has died. While respirators and other medical technology and efforts are able to maintain many bodily functions and processes, their continued employment is considered medically inappropriate even if the patient's family or other surrogates might want them continued. While physicians are and ought to be sensitive to the needs of families to adjust to the death of a patient where the criterion of death is brain death and the patient may retain much of the appearance of the living, families are quite properly no longer presented with decisions about continuing medical measures once death has been declared. The dead no longer have any moral claims to health care grounded in justice.

This same conclusion follows whether one approaches the question from either the prudential allocator or interpersonal distribution standpoint. On all common views of personal identity, when we die *we* cease to exist and only our bodies remain. Since health care cannot anymore serve *our* good in any way by affecting our opportunity, suffering, or lifespan – how could it when we have ceased to exist? – we have no reason from the intrapersonal standpoint of the prudential allocator to want to continue such care for our bodies. Nor do others from an interpersonal standpoint have any reason to suppose that any such dead body that is not their own can be benefitted in any way by health care.

III. THE JUST CLAIMS TO HEALTH CARE OF THE PERSISTENT VEGETATIVE STATE PATIENT

Consider now a second kind of patient who is commonly assumed to be in many respects the closest approximation to the patient who has died – the patient now in a persistent vegetative state, having suffered an irreversible and complete loss of consciousness.[7] This patient differs from the patient who has died in that the PVS patient retains some lower brain activity or brain stem function permitting continuation of many self-regulating bodily functions. However, with the irreversible loss of consciousness all capacity for experience – hopes and fears, joys and sorrows, pleasures and pains, plans and purposes – has now been irreversibly lost, and with it as well all capacity for purposive action or agency. As already noted, PVS patients will include both some end-stage dementia patients as well as other patients who have suffered severe cerebral trauma and insult. Can any of the aims of medicine still be pursued and achieved for such patients? The promotion of opportunity through the prevention or treatment of disability is no longer possible since disability is irreversible and total; opportunity presupposes capacities for agency and action. The prevention or palliation of pain and suffering is no longer possible because the capacity for pain and suffering presupposes the consciousness that has been irreversibly lost. The extension of the life of the human being does remain possible so long as we grant that the relevant criterion for the death of the human being is the *whole* brain criterion. However, does the continuation of the life of the human being who has suffered an irreversible loss of consciousness still serve the interest of the individual who is that human being?

I will distinguish two plausible ways of thinking about this question, both of which yield a negative answer: the first holds that the *person* has died or has ceased to exist, and so no treatment of the human being could now further the person's interests; the second seeks to show that, whether person or merely human being, a PVS patient has no interest in life-sustaining treatment as such. The first line of argument will be taken by those who accept the so-called higher brain version of brain death as a criterion for the death of the person. They will conclude that the person whom this patient was has ceased to exist or has died. No health care could possibly extend further the life of the *person*. Suppose we then grant that the life of the person has ended with the permanent loss of all capacities for consciousness, while the life of the human being or human biological organism, for example as reflected in legal statutes concerning death, continues. The question then shifts to the second line of argument concerning whether the aim of medicine of preserving and extending life should be understood as applying only to persons or also to human beings in a persistent vegetative state who are no longer and never again will be persons.

I believe one helpful way of thinking about this issue is to consider the question in one's own case. If you were to suffer such severe brain damage as to leave you in a persistent vegetative state, would you still have an interest in measures aimed at sustaining or extending your life, or would continued life-sustaining measures no longer be of any interest or benefit to you? I believe that nearly all persons who reflect on this question conclude that, because the complete and permanent loss of consciousness carries with it the loss of all possibilities for agency and experience of any sort, continuing to sustain the life of the patient's body is of no benefit whatever to the patient. However the complex notion of interests is analyzed, a thing's having interests of its own in some object or state of affairs x is usually tied to the thing's present or future capacity for sentience and to its capacity to care about x.[8] PVS patients have permanently lost all capacity to care about or take an interest in anything. So far as their present or future capacities for any psychological life or conscious experience are concerned, PVS patients are similar to such living things as plants, which lack any capacity for sentience or a conscious life; they lack the sentience that even many animals have.

It would appear then that the only self-regarding interest a person might have in continued life-sustaining treatment after having suffered a complete and irreversible loss of consciousness is based on the possibility that the diagnosis and prognosis are mistaken and that the condition is not irreversible. However, as the probability that the diagnosis is mistaken becomes vanishingly small and the condition of the patient in the very rare cases of recovery is one of severe disability, any interest of the patient in continued life-sustaining care becomes vanishingly small as well. On the other hand, the PVS patient has also lost all capacity to experience any burdens or suffering from continued treatment, or from continued life in general. Consequently, life-sustaining treatment for PVS patients cannot plausibly be held to impose greater burdens, in the form of burdensome *experience*, than benefits on the patient. Sometimes treatment will be unwarranted because it is known that the patient did not want, or would not have wanted, treatment in these circumstances because she believed either that the treatment would be of no benefit to her or that it would harm or adversely affect the interests of others about whom she cared. In such cases, the patient can be harmed by continued treatment in the sense of having her earlier desires frustrated and her interests adversely affected, though she cannot be hurt in the sense of undergoing any hurtful experience. In other cases treatment may be unwarranted because the extremely small expected benefit to the patient from continued treatment is clearly and substantially outweighed by the burdens or costs to others, such as the patient's family, of that treatment.

More to the point here, however, is that neither the prudential allocator nor the interpersonal distribution approach seems able to ground claims of the PVS patient to any life-sustaining care for the benefit of the patient. A prudential allocator deciding whether to make provision for the possibility of life-sustaining care in such circumstances would almost certainly prefer to use the necessary resources on other desired and valued activities that could contribute to his or her life prior to suffering the (virtually certain) permanent and complete loss of consciousness. The possibility of error in the diagnosis is too small, and the patient's likely quality of life even should the diagnosis be mistaken too severely compromised, to make the expected benefit of continued care warrant its cost in comparison with other possible uses of those resources. If costs are not taken into account, a person might reason that even a vanishingly small possibility of prognostic error could justify from the patient's point of view employing life-sustaining treatment, since the patient can experience no burdens. However, the prudential allocator approach shows that this reasoning will not ground any claim on social resources to fund the treatment. This result fits our judgments and practices in other cases. It is not considered any requirement of justice to continue to search for those lost at sea or trapped in mines so long as there is any possibility of saving them, no matter how infinitesimally small.

On the alternative interpersonal distribution approach in which obligations are owed to persons, it is implausible to maintain that a person still exists in the case of a human being who has suffered a complete and irreversible loss of consciousness. Philosophical analyses of the concept of a person are notoriously difficult and controversial, yet I believe the main alternative accounts share the assumption that personhood is incompatible with the complete absence of *any* present or future capacity for purposive agency, social interaction, or conscious experience of any sort whatever. Human beings who have suffered this tragic loss lack even the capacities for pleasure and pain, and for goal-directed action or behavior, of animals that are uncontroversially held to lack the capacities necessary for personhood.

It might mistakenly be thought that the philosophical literature on personal identity shows that whether the capacity for consciousness is necessary to personhood is at the least controversial. The possible confusion here is instructive. There are two broad positions about the criteria for personal identity in the philosophical literature on that subject. One takes identity over time to be established by physical or bodily continuity between different stages of the person, while the other locates identity in the continuity and connections of psychological states of the person over time.[9] (In fact, there is a third broad position on the criteria for personal identity supported by some philosophers that holds both physical *and* psychological continuity to be necessary for identity. I will ignore this alternative in what follows since I do not believe it affects my argument here.)[10] If one of the two principal contenders for the correct theory of personal identity appeals to physical continuity, how then can it be uncontroversial that consciousness, plainly a psychological state, is necessary to personhood?

Here, it is important to distinguish an analysis of the concept of a *person* from criteria for personal *identity* over time. Theories of personal identity are constructed to determine when one person existing at a particular time is identical with, that is, is the same person as, a person existing at another time. Proposed criteria of personal identity are tested in the philosophical literature by various hypothetical cases, often of a very bizarre and science fiction nature, involving fission, disintegration and reformation at different times and places, and so forth. What must be emphasized is that physical or psychological continuity are proposed as criteria for the *identity* of persons over time, not as criteria of personhood. Their adequacy is tested by cases that elicit and test our beliefs about whether two persons that exist at different times are the *same* person, not whether an entity existing at one of the two times is a person at all. The cases are commonly such that it is uncontroversial that the different entities that may or may not be the same person, are in each case plainly persons. Thus, when physical continuity is proposed as a criterion for personal identity over time, it is commonly presupposed that the two or more entities in question each possess whatever properties or capacities of a psychological sort are necessary for personhood itself.

To take a simple example, suppose that by some process of fission Smith divides into two distinct individuals, Smith-2 and Smith-3. Smith is physically continuous with Smith-2 *and* Smith-3 and, let us further suppose, Smith-2 has exactly the same physical properties as Smith-3 (except for relational properties such as spatial location). On the physical continuity account of identity, presumably Smith is identical with *both* Smith-2 and Smith-3. If Smith has the same or connected beliefs, memories, and so forth as Smith-2, whereas Smith-3 has an entirely different and unconnected set of beliefs and memories from Smith (setting aside problems of whether it is possible to have different psychological properties without

367

attendant differences in physical properties), then on the psychological account of identity, Smith is identical with Smith-2 but not with Smith-3.

But now let us suppose that our master scientists carrying out these amazing fission procedures have not yet fully perfected the process and that Smith-3, while still physically continuous with Smith, has now had his brain so damaged that all its functional capacities have been irreversibly and completely destroyed, including its capacity to support consciousness. While physical continuity remains largely intact, the physical continuity identity theorist will no longer claim that Smith and Smith-3 are the same person; but neither will he claim that Smith-3 is a different person than Smith. Rather, Smith-3 is unfortunately no longer a person at all since the botched procedure destroyed all functional capacities of the brain and with it consciousness. Smith-3 has died, by the standard whole brain criterion for death; while Smith-3's body remains, the person was destroyed and now no longer exists. This is no different than the ordinary case in which a person dies. There too, the physical continuity identity theorist holds that the person no longer exists and only his body remains.

Thus, I believe some degree of present or future capacity for sentience, conscious experience, or purposive behavior is uncontroversially necessary for being a person in the sense relevant to being an independent source of claims within theories of distributive justice. Whether or not we claim that the PVS patient's irreversible loss of consciousness satisfies the proper brain death criterion for the death of the patient or human being, it does make highly implausible any claim that patient might be thought to have on social resources for the provision of health care, within either the prudential allocator or interpersonal distribution approaches to the determination of those claims. Even granting that human life, if not a person, still exists in a PVS patient, both the prudential allocator and interpersonal distribution approaches show why the nature of that life is not itself such as to ground claims to health care.

It might be claimed that a PVS patient who was in the past self-conscious and uncontroversially a person should still be considered a person until the death of the body or of the human being, despite having suffered an irreversible and complete loss of consciousness. Karen Quinlan, for example, is considered not to have died until some nine years after entering a persistent vegetative state. Ronald Dworkin (1986) has argued that if we take personal identity to consist in a classification of the facts about psychological and/or physical continuity, and not some further deeper and independent fact to be discovered, then it is a matter for social and moral decision whether a PVS patient satisfies the necessary continuity to remain a person. Dworkin goes on to argue that this decision should be made in a manner that is most consistent with our other moral, social, and legal practices, most of which seem to presuppose that the person continues to exist until the death of the body or the human being.

Two points need to be made in brief response to this line of argument. The first is that it too seems to conflate whether the PVS patient is the *same* person as existed earlier before entering the persistent vegetative state, with whether the PVS patient is still a person *at all*. It may be morally and socially undesirable to have disconnected lives, as Dworkin claims, but that speaks only to whether, if the PVS patient remains a person, we should hold him or her to be the same person. The question at issue, however, is whether the PVS patient is any longer a person at all. The widespread acceptance of the brain death criterion for the death of a human being in effect already rejects the death of the body as marking the death of the person, and my suggestion is that the irreversible loss of all capacity for any conscious experience of the PVS patient warrants distinguishing between the death of the person (so-called higher brain death) and the death of the human being (so-called whole brain death). The other point is that the concept of a person relevant here is for a theory of justice in health care: persons as possible sources of claims to health care grounded in justice. If PVS patients cannot have any possible interest in health care, then it is unclear how they could be persons in the sense of possible independent sources of claims to health care.

IV. THE JUST CLAIMS TO HEALTH CARE OF THE SEVERELY DEMENTED PATIENT

As I shall understand them here, severely demented patients, including patients with end-stage or near end-stage dementia who are not in a persistent vegetative state, remain subjects of conscious experience, and so remain capable of suffering pain and sometimes as well of enjoying at least some simple pleasures. These patients are plainly still live human beings by any standard. Moreover, they do not lack *all* the capacity for experience and agency that make it plausible to hold that PVS patients are no longer persons. Nevertheless, in patients with severe irreversible dementia cognitive capacities and, in particular, memory are so destroyed as to eliminate or drastically reduce the psychological continuity and connectedness on which personal identity, according to the psychological continuity theory of identity, rests. Like PVS patients, the severely demented are patients whom family members often think of as now being only the physical shells of the persons they once loved, the persons now having been destroyed by the ravages of the dementia. We must address the question of what such patients are owed in the way of health care both properly mindful of the persons they once were and of the web of relationships in which they existed, while at the same time squarely facing the tragic state to which their dementia has brought them.

With the severely demented patient, the two main alternative theories of personal identity appear to have sharply different implications. Physical continuity of the body clearly is maintained so that on that theory of identity, if a person still exists, it is the same person as had earlier existed. Moreover, unlike the PVS patient, there is still some capacity for conscious experience, though without the psychological experience in the patient of continuity over time, looking both back over time as well as forward into the future, that normal adult humans possess. The radical destruction of memory breaks the patient's felt or experienced links with his or her past as well as prevents the establishment of new links with the future as it becomes present and then past experience. The "temporal glue" that memory contributes to psychological continuity over time is lost and so personal identity on the psychological continuity theory would seem either no longer to exist or to be substantially undermined.

Is it the case that personal *identity* is undermined, but as with fission and fusion type cases a person, though perhaps only not the same person, still exists, or is it the case that as with the PVS patient it is more plausible to hold not merely that the *same* person no longer exists, but that personhood itself has been destroyed and *no* person now exists? Are families correct in believing that with a severely demented patient, as with a PVS patient, only the body, the shell of the person that once existed, now remains? I believe this question of personhood is exceedingly difficult, raising some of the deepest and most complex issues of philosophy. Quite obviously, we cannot even address, much less settle, all of these complexities in the space available here. All I can do is indicate the direction in which I believe such an inquiry would lead.

To help get some handle on this issue of the personhood of the severely demented, we might consider how we would or should regard beings that *never* had advanced beyond the mental life of the severely demented. While wishing not to offend those who care deeply for the persons who have now become severely demented, I believe the comparison with some animals is instructive. It is widely agreed, for example, that dogs and horses are not persons. Presumably, this is not simply because they are not members of the human species, since it is at least possible that there are non-human persons, but because they lack some important properties or capacities that humans normally possess.[11] Animals such as dogs and horses surely are sentient beings – they are conscious and capable of experiencing pleasure and pain, and in particular they can be made to suffer. What they presumably lack is the capacity for or experience of *self-consciousness*, a conception of themselves as, and experience of being, a single self-conscious individual who persists through time. Thus, while they can experience pain and suffer here and now in the present, and can be conditioned to associate pain and suffering with experiences not themselves painful, they lack capacities for hopes and fears, dreads and longings for their futures, or more generally the capacity to form plans for their future. For this, they would require what it is commonly believed they lack, a belief that they themselves are beings that persist through time with a continuity of self-consciousness over time.[12]

This may go some way toward explaining what otherwise appears an anomalous feature of many persons' moral views about the treatment of animals. Many persons hold that causing gratuitous pain or suffering to animals such as dogs and monkeys is seriously wrong though painlessly killing such animals is not wrong. The apparent anomaly is that with persons it is commonly believed that killing them against their will is one of the most serious wrongs, if not *the* most serious wrong, that can be done to them, and specifically is a more serious wrong than causing them pain or suffering. Why then for animals is the more serious wrong of killing commonly held not to be a wrong at all, while causing them pain or suffering remains a wrong? I believe the explanation lies in the differences between humans and animals that I have just noted – while each can be caused pain, which is immediately experienced as unpleasant and unwanted, only humans but not animals can have plans and desires about their future, and indeed have desires that they have or experience that possible future, which can all be frustrated, or at least left unsatisfied, by being killed.[13] It is this capacity to envisage and desire a future for oneself that I believe best explains why killing a normal adult human wrongs that person.[14]

I believe that the severely demented, while of course remaining members of the human species, approach more closely the condition of animals than normal adult humans in their psychological capacities. In some respects the severely demented are even worse off than animals such as dogs and horses, who have a capacity for integrated and goal-directed behavior that the severely demented substantially lack. The dementia that destroys memory in the severely demented destroys their psychological capacities to forge links across time that establish a sense of personal identity across time. Hence, they lack personhood. This means in turn that they lose the fundamental basis for persons' interest in continued life and in measures that sustain life – that their future life is a necessary condition for satisfying all of a person's desires about and plans for the future.

It may seem that this is merely a judgment about the quality of life of the severely demented, and as such subject to the common worries about making such judgments about others. However, I believe it is not merely a judgment that the severely demented suffer a tragically diminished quality of life, though at least that much is of course true. Instead, the point is that the severely demented have been cut off from the self-conscious psychological continuity with their past and future that is the basis for the sense of personal identity through time and which is a necessary condition of personhood. They lose the capacity for past and future selves all being a single persisting "me." As best others can determine, their experience becomes a disconnected or disjointed blur, incoherent and so to them essentially unintelligible. Personhood involves criteria of psychological continuity that are not satisfied in the severely demented.

As with the dead patient and the PVS patient, physical continuity is not sufficient, in the absence of psychological continuity, to establish that the patient is still a person. The difficulty with the physical continuity criterion drawn from the personal identity literature is that generally it is used there to establish the identity or sameness of persons at different times when there is a physical continuity between the two person stages existing at those times, but a psychological discontinuity between the person stages existing at those different times, *and also* at least some psychological continuity within the person stages on either side of the psychological break in continuity. For either a dead patient or a PVS patient, physical continuity fails to establish personal identity at two points in time because all capacity for psychological and conscious experience of the person who had occupied that body at the earlier time while self-conscious has irreversibly ceased at the later time; neither dead patients nor PVS patients remain persons. A present or future capacity for conscious experience is a necessary condition for personhood itself in the later body, whatever more may or may not be necessary for personal identity over time.

Now *this* reason for denying personhood at all to the dead or PVS patient does not justify a similar denial of personhood to the severely demented patient in whom some capacity to experience at least primitive pleasures and pain continues to exist. However, severely demented patients do lack any attendant capacity to build psychological connections back into the past or forward into the future; they cannot establish an experienced continuity of a single person who experiences the different pleasures and pains over time. In the extreme, the relation between the experience of the severely demented patient who experiences some pain today and another tomorrow is as alien as that between two experiences of pain of two *different* normal humans. Just as the animal that lacks a conception of itself as a single continuing subject of experience over time, and as a result necessarily lacks desires, hopes, and so forth, about its future, is for that reason plausibly held not to be a person, the severely demented patient is likewise plausibly held not any longer to be a person because he or she too lacks all psychological continuity and connectedness over time. Some psychological continuity and connectedness is necessary for personhood itself, before questions of the identity of persons existing at different times might be raised and possibly settled by continuity criteria of either a psychological or physical sort.

The claim that patients who are severely demented, but who have not suffered an irreversible loss of consciousness and so remain capable of some experience of pleasure and pain, are no longer persons is a controversial one and there is not space either to develop or to defend the position fully here. But it is at least possible to avoid some potential confusions about the nature of the claim that may make it seem less plausible and more controversial than I believe it is.

First, it does not follow from the claim that the severely demented are no longer persons that they therefore lack *any* moral status or *any* valid claims on grounds of justice to health care treatment. Having lost the psychological continuity necessary to personhood, they have lost the capacity to envisage a future for themselves and in turn the capacity to have desires about, or for, that future. I believe this implies that the severely demented have lost an interest in treatment whose ultimate purpose is to prolong or sustain their lives. But since, unlike the dead and PVS patients, they remain sentient and capable of conscious experience, they retain an interest in the quality of that experience so long as they remain alive and in their present condition. In particular, they retain an interest in receiving whatever health care treatments and other measures may relieve or palliate any pain or suffering of a physical, psychological, or emotional sort they might otherwise undergo, as well as an interest in receiving any measures that may provide some pleasures in their lives.[15] However, this is an interest in receiving pleasure under the assumption that, or while, they continue to live, but not an interest in continuing to live in order to receive any pleasures that might be possible for them. Just as there is no serious moral reason to bring non-person beings, such as animals, into existence simply for the sake of the pleasure they might have, there is no reason to prolong the life of non-persons, whether animals or human beings, for the sake of the pleasure they might have. This parallels the position that has been taken in population policy discussions that there is a moral reason to seek to make existing people happy or provide people with pleasure, but not a moral reason to create people or other beings because they will be happy or have pleasure.[16]

Like all patients, the severely demented also retain an interest in care, such as simple measures to keep them clean, that treats them with dignity out of respect for the person they once were. Indeed, it is generally acknowledged that the bodies even of the dead must be treated with dignity respecting the person, and the person's wishes, whose body this once was.

The second point of clarification is that I recognize a relatively normal person who is told that he suffers from early Alzheimer's that will inevitably progress to the stage of severe dementia would likely identify that severely demented future individual as himself. If asked to assume the standpoint of a prudential allocator, the person would likely express what he understood to be a prudential concern for the well-being of the severely demented patient with whom he shares at least physical continuity. In particular, the expected future suffering that the later severely demented patient will undergo would likely be viewed now by this person with early Alzheimer's, not with compassion for *another's* suffering, but with foreboding of the future suffering that *he* will experience. Moreover, others who knew and cared for the person before he or she was ravaged by dementia would likely identify this patient to be the same person as the earlier person with whom the patient shares physical continuity.

Do these facts suggest that my argument above was mistaken and that personal identity is maintained by physical or bodily continuity while psychological continuity is not necessary for the severely demented patient either to continue to be the same person as existed in that body before the dementia or to continue to be a person at all? I believe it does not. Some would dismiss this concern for the severely demented patient one will become as an irrational fear based on a mistaken identification with what happens to our bodies or our "successors." An alternative is that an individual *human being's* identity can extend beyond the limits of his or her *personal* identity, and indeed beyond the limits of the continued existence of any person *at all* still "residing" in that human being's body. If I now envisage the prospect in the future of becoming severely demented and then undergoing painful suffering, it is consistent and plausible to assume that physical continuity, in particular continuity of the brain, establishes some sense in which *I* will experience that future suffering, though I will no longer be either the same person that I now am nor any person at all. It is as if I had been transformed, with respect to my psychological capacities for conscious life, into an animal of the sort that uncontroversially lacks personhood, though capable of experiencing some pains and pleasures.

If this is correct, it tells us something important about the concerns of the prudential allocator considering how to allocate his or her fair share of resources to health care. Our earlier claim that the concerns of the prudential allocator were bounded by the limits of the personal identity of the allocator was mistaken and requires amendment. When the physical continuity of the brain that supports consciousness is maintained, a person's prudential concern can extend beyond the limits of the individual's personal identity or even the limits of the continued existence of the organism as a person. Indeed, we might already have noticed this general conclusion in considering questions about the treatment of the body of a person who has died. Persons do often express what I believe is a form of prudential concern that their bodily remains after death be treated and disposed of with dignity and respect.[17] In sum, the change in the condition of the patient suffering severe dementia does destroy the patient's personhood, but instead of eliminating the possibility of prudential concern by the earlier person for the non-person he or she will later become, this change in condition alters the nature of the concerns that the prudential allocator will have for the severely demented patient he or she will become.

With regard to health care, the alteration will be in what health care a person will have a prudential concern to provide should he or she later suffer severe dementia – the concern will be with palliative care directed at relieving or preventing suffering and maintaining comfort, and not with life-sustaining care. And this means that if the prudential allocator approach is sound for determining the just claims to

377

health care, the access to which a society is ethically obligated to insure to all its members, then a crucial distinction would be made in the health care owed to the severely demented between life-sustaining and -prolonging care as opposed to palliative and comfort care. While I will not spell out the argument, the interpersonal distribution approach has essentially the same implications. The ends of promoting persons' opportunity or prolonging their life can no longer be achieved for the severely demented. Justice in health care for the severely demented requires access only to palliative care that improves comfort and relieves suffering, though it is possible that even some palliative measures might be so expensive that prudential allocators would limit the resources they would devote to them.

V. THE JUST CLAIMS TO HEALTH CARE OF THE MODERATELY DEMENTED PATIENT

I turn now to consider very briefly the just claims to health care of the moderately demented patient. The effects of dementia are various and many, and the degree to which any one effect is present in a given patient varies continuously along what is commonly a broad spectrum. Thus, any classification of patients of the sort I am employing here into severely or moderately demented is inevitably very crude and misses important differences between real patients. Nevertheless, I think the distinction can be useful in pointing to an ethically important difference in the condition of demented patients. I characterized above some end-stage demented patients as having suffered an irreversible loss of all consciousness and so as being in a persistent vegetative state, while severely demented patients remain conscious but have suffered such drastic loss of memory that their personal identity has been substantially destroyed. I shall understand the moderately demented as at least still having enough memory function and/or other psychological continuity to maintain personal identity (or to maintain it to a substantial degree if personal identity can be maintained in degrees), though as also often suffering from cognitive disabilities sufficient to leave them incompetent to make at least some decisions about their health care.

Some of these patients will be what are sometimes called the pleasantly senile or demented who exhibit little evidence of subjective distress or suffering from their dementia. In other cases, the effects of dementia on patients in this broad category may leave them angry, confused, depressed, and suffering from substantial subjective distress as a result of their dementia. There is thus a wide variation in the moderately demented regarding the extent to which their condition appears now to be experienced by them as on balance distressful or pleasant. There is a second ethically important range of variation concerning the moderately demented that is distinct from differences in their condition when moderately demented – this is the difference in individual reactions to and evaluations of the prospect of becoming moderately demented, and specifically to whether one would or would not then want particular forms of life-sustaining treatment for oneself. The increase in suicide rates for moderately demented patients, as well as the testimony of competent persons who have had experience with the moderately demented about what they would want for themselves in such circumstances, both suggest that at least some significant numbers of persons, though by no means all persons, find the continued life of the moderately demented that is sometimes supported by various kinds of life-sustaining treatment on balance no longer a benefit or desirable in comparison with the alternative of an earlier death.

This would then be life-sustaining health care that some prudential allocators would not wish to include in their insurance package simply because it would not be wanted even apart from its cost. For other prudential allocators, while the life of a moderately demented patient supported by some forms of life-sustaining treatment would not be judged to be on balance worse than no life at all, the cost of such care should they become moderately demented may be such that they would choose to use those resources for other purposes while not demented. They would choose not to include this care in their insurance package. Still other prudential allocators, however, would wish to have, and so would insure for, relatively non-burdensome life-sustaining care for themselves should they become moderately demented and pleasantly senile.

This difference in the judgments that prudential allocators would make about provision for life-sustaining care for the moderately demented means that the prudential allocator model is inconclusive regarding whether general social policy covering all members of society should include or exclude this health care from insurance coverage. Institutional mechanisms that actually permitted real persons with fair income shares to accept or reject insurance coverage for such care would reflect this variation and seem not to be unjust. However, so long as the insurance coverage for such care is determined largely as a matter of general social policy, with one standard generally applying to all, as is the case for health insurance coverage for the elderly under Medicare, it probably is more just to cover life-sustaining care for the moderately demented. Individual variation then would be reflected in decisions in accordance with the wishes of patients while competent, or with whether treatment would now be found unduly burdensome by the moderately demented patient. I stress that my conclusion here is that life-sustaining care should not be denied patients under programs like Medicare or Medicaid *merely because* they are moderately demented. It may be, however, that age itself, for example whether persons are below or beyond the normal lifespan, is relevant to their just claims to life-sustaining care, but that is another matter which I do not take up here. My concern in this essay is only with the relevance of dementia and cognitive disability to just claims to health care.

On the other hand, I believe it is clear that virtually all prudential allocators would want to provide for care aimed at increasing their comfort and relieving pain or suffering at not great cost should they become moderately demented, especially under present circumstances in which voluntary active euthanasia is legally impermissible. This care should be part of any adequate level of health care socially guaranteed for all, and should not be denied patients simply on grounds of cost. There are, however, often formidable practical difficulties in distinguishing life-sustaining from palliative care, and sometimes particular health care inextricably serves both ends.

In conclusion, I want to underline two aspects of the limited scope of my discussion here. I have tried to sketch how and why different levels of dementia and associated serious cognitive disability suffered disproportionately by the elderly affects their just claims to health care under both prudential allocator and interpersonal distribution approaches to that question. One very important set of concerns that there has not been space to address here at all is the symbolic importance of how our society cares for these patients and how, in particular, limiting life-sustaining care for them might lead to *un*justified withholding or withdrawal of life-sustaining care for other patients, together with a generally undesirable deterioration in our society's care and concern for its frail and vulnerable members. These are real worries that certainly deserve to be seriously addressed before concluding that the position I have sketched above concerning the health care owed to the seriously cognitively disabled on grounds of justice should be translated into health policy. The second limitation concerns how, or even whether, this ethical position might be translated into health law and health policy in the United States in a socially and politically acceptable way, for example, through changes in Medicare reimbursement policies.[18] Both of these concerns about what I have not addressed here will only loom as important, however, if the ethical position I have sketched out is otherwise plausible. I am confident in any event that the issues need more public discussion than they have as yet received.

NOTES

1   Such a view is developed by Gibbard (1983) and Daniels (1983),

and is elaborated in more detail in Daniels (1988). See also Dworkin (1986).

2   I discuss the moral relevance of these factors in Chapter 9 of this volume.

3   This prudential allocator approach makes assumptions about the nature of personal identity that have been powerfully challenged by Parfit (1984). If Parfit is correct, personal identity is a matter of degree, not an all or nothing matter, and does not have the importance that the prudential allocator approach seems to assign to it. Nor, on Parfit's view, are the boundaries around the life of a single person at the places that approach assumes. There is space here only to note, but not to explore, the importance of Parfit's work to these issues.

4   A representative summary account of this sort can be found in President's Commission for the Study of Ethical Problems in Medicine (1983b). An excellent survey of philosophical work on access to health care is Wikler (1983).

5   The theoretical controversy centers largely on whether the whole brain definition should be replaced by the so-called higher brain definition so that patients would be determined to be dead when all upper brain functions that support cognitive activity and consciousness had irreversibly ceased, although some brain stem function remained. For a defense of the whole brain definition, see President's Commission for the Study of Ethical Problems in Medicine (1981). For defenses of the higher brain formulation, see Veatch (1975) and Green and Wikler (1980).

6   Persons might have one interest that extends beyond respectful treatment of their bodies after they have died that could ground claims to health care after they have died. This is the interest in maintaining bodily functions to preserve organs for donation to others or to allow for research or experimentation on their bodies. This health care, however, primarily serves the interests of others – the living recipients of the organs or the beneficiaries of the research – not the person who has died. My interest in this paper is with the claims to health care of the severely demented in their own right and not simply for the benefit of others. I will not repeat this qualification about a possible interest in benefitting others with one's body, but it is implicitly assumed at several places in my discussion below.

7 For a general discussion of decisions about life-sustaining treatment for PVS patients, see President's Commission for the Study of Ethical Problems in Medicine (1983a, ch. 5).

8 For an excellent discussion of the complexities in the concept of an interest, see Feinberg (1984, ch. 1).

9 Two helpful sources for the philosophical literature on both personhood and personal identity are Rorty (1976) and Perry (1975).

10 All three of these theories of identity locate the criteria of identity in non-moral facts about the person. Some philosophers, e.g., Daniels (1979) and Rorty (1976, pp. 3–4), have held a moral/social account of personal identity according to which it is established, at least in part, by moral considerations, or by "non-personal facts" such as others' responses to and judgments about the person. Some such alternative position seems clearly correct concerning broader or "thicker" conceptions of the person that are parts of normative moral theories like utilitarianism and contractarianism. My concern here is with "thinner" conceptions of personal identity and personhood that identify what makes us beings who have, and know that we have, a future. I believe it is the criteria for these thinner conceptions that ground our interests both in continued life and in self-determination. I take these interests to be in turn necessary to our interests in and claims to life-sustaining and opportunity-promoting health care; as I note later in the text, they are not necessary to our interest in merely palliative care. I do not have space here to explore any of these moral/social conceptions of personal identity, and so my argument is limited to the standard psychological or physical continuity conceptions. The alternative moral/social conceptions make clear that conceptions of personal identity, personhood, and the person play a number of distinct theoretical roles, and it is a mistake to assume that a single account is correct in all contexts.

11 The classic statement of the "speciesism" argument is in Singer (1975).

12 Animals like dogs do appear to anticipate at least near-term future events, such as the return of their owner or being fed. It is a complicated matter, however, how such behavior should be interpreted and to what extent, if any, it supports attributing to them a conception of themselves as beings with a conception

of a continuous conscious self over time. For a discussion of the complexities of such attributions in the (perhaps) easier case of human infants, see Tooley (1983, ch. 11).

13 Michael Tooley (1983), among others, has supported this moral difference in the status of persons and many animals in roughly this way. However, Tooley's argument that beings are not persons, and do not have a right not to be killed, if they permanently lack a capacity to have a conception of themselves as temporally continuous subjects of experience and other mental states, is different from mine in a respect that is worth noting. While he has formulated his position in different ways on different occasions, in each case what is asserted is a logical or conceptual relation between believing oneself to be a continuing self and having a right not to be killed. For example, in the earliest formulation (1972), he claimed that it is a conceptual feature of rights that they can be violated, but that an action that deprives A of x does not violate A's right to x if A does not desire x. Consequently, A must have the capacity to desire x in order to be the kind of being that could possibly have a right to x. In some form, I believe that such a logical claim is probably sound, though, as Tooley is well aware, the argument is much more complicated than I have indicated here.

There is an alternative interpretation of this argument, however, which is compatible with accepting the argument in its logical form as well, that construes it in an explicitly normative or moral form. Put loosely, the idea of the moral version of the argument locates why it is seriously wrong to kill persons in the fact that persons can and commonly do have desires to experience their future, and have many desires about or for their future. Continued life is a necessary condition for the satisfaction of any of those desires, and being killed frustrates all of them. Animals unable to form a conception of themselves as temporally continuous subjects of experience, on the other hand, cannot have desires for or about their future, and so are not denied anything about which they care in being killed. A future can thus have value for a person in a way that it cannot for an animal that is not a person.

Of course, a great deal more must be said in elaborating and defending either version of the argument and, as already noted, I do not believe that one must choose between the logical and moral interpretations of the argument. I do believe, however, that the moral version has considerable force independent of the logical version.

14 I acknowledge, though largely only to set it aside here, that my view of personhood implies that infanticide need not wrong a newborn infant and that infants lack any serious moral right not to be killed. Like Tooley (1983), I believe the widespread agreement in our society that infanticide is wrong is not in itself sufficient to establish that my position on personhood is mistaken. That infanticide seriously wrongs a newborn needs persuasive argument and defense, which I believe proves remarkably difficult to provide. There remain, of course, other serious moral reasons against infanticide, either as a general social practice or for the vast majority of individual infants, that do not rest on a wrong done to the infant. Whether the implications of my conception of personhood for both infanticide and abortion are acceptable is another respect in which my discussion in this essay is incomplete.

15 The importance for physicians and others caring for the severely demented of clarity about the goals of care is stressed in an excellent paper by Rango (1985).

16 This is essentially what Jan Narveson (1976) calls the person affecting restriction; the restriction is criticized in Parfit (1984, ch. 18). It is noteworthy that even utilitarians need not reject my point in the text if they accept the person affecting restriction.

17 Allen Buchanan (1987) has distinguished a right to self-determination that persons have regarding what happens to them while still a person from a right of disposal concerning what happens to their body or property when they are no longer persons. In these terms, my suggestion is that the concern of the prudential allocator extends beyond the actions covered by a right to self-determination to those covered by a right of disposal.

18 For an interesting exploration of some implications of personal identity for appropriate standards for health care treatment decision-making for incompetent patients, see Dresser (1987).

REFERENCES

Brock, D.: 1986, 'The value of prolonging human life', *Philosophical Studies*, 50, 401–428.

Buchanan, A.: 1987, 'Our treatment of incompetents', in T. Regan and D. VanDeVeer (eds.), *Health Care Ethics*, Temple University Press, Philadelphia.

Buchanan, A., and Brock, D.: 1986, 'Deciding for others', *The Milbank Quarterly* 64, Suppl. 2, 17–94.

Daniels, N.: 1979, 'Moral theory and the plasticity of persons', *The Monist*, 62, 265–287.

Daniels, N.: 1983, 'Am I my parents' keeper?' in *Securing Access to Health Care, Volume Two: Appendices. Sociocultural and Philosophical Studies*, Report of the President's Commission for the Study of Ethical Problems in Medicine, U.S. Government Printing Office, Washington, D.C.

Daniels, N.: 1988, *Am I My Parents' Keeper?: An Essay on Justice Between the Young and the Old*, Oxford University Press, New York.

Dresser, R.: 1987, 'Life, death, and incompetent patients: conceptual infirmities and hidden values in the law', *Arizona Law Review*, 28 (3), 373–405.

Dworkin, R.: 1986, 'Philosophical issues concerning the rights of patients suffering serious permanent dementia', prepared for the Office of Technology Assessment, U.S. Congress.

Feinberg, J.: 1984, *Harm to Others*, Oxford University Press, Oxford.

Gibbard, A.: 1983, 'The prospective Pareto principle and equity of access to health care', in *Securing Access to Health Care. Volume Two: Appendices. Sociocultural and Philosophical Studies*, Report of the President's Commission for the Study of Ethical Problems in Medicine, U.S. Government Printing Office, Washington, D.C.

Green, M., and Wikler, D.: 1980, 'Brain death and personal identity', *Philosophy and Public Affairs*, 9, 105–133.

Narveson, J.: 1976, 'Moral problems of population', in Michael Bayles (ed.), *Ethics and Population*, Schenkman, Cambridge, Massachusetts.

Parfit, D.: 1984, *Reasons and Persons*, Oxford University Press, Oxford.

Perry, J. (ed.): 1975, *Personal Identity*, University of California Press, Berkeley.

President's Commission for the Study of Ethical Problems in Medicine: 1981, *Defining Death*, U.S. Government Printing Office, Washington, D.C.

President's Commission for the Study of Ethical Problems in Medicine: 1983a, *Deciding to Forego Life-Sustaining Treatment*, U.S. Government Printing Office, Washington, D.C.

President's Commission for the Study of Ethical Problems in Medicine: 1983b, *Securing Access to Health Care*, U.S. Government Printing Office, Washington, D.C.

Rango, N.: 1985, 'The nursing home resident with dementia', *Annals of Internal Medicine*, 102, 835–841.

Rorty, A. (ed.): 1976, *The Identities of Persons*, University of California Press, Berkeley.

Singer, P.: 1975, *Animal Liberation*, A New York Review Book, New York.

Tooley, M.: 1972, 'Abortion and infanticide', *Philosophy and Public Affairs*, 2, 1, 37–65.

Tooley, M.: 1983, *Abortion and Infanticide*, Oxford University Press, Oxford.

Veatch, R.: 1975, 'The whole-brain oriented concept of death: an out-moded philosophical formulation', *Journal of Thanatology*, 3, 13.

Wikler, D.: 1983, 'Philosophical perspectives on access to health care: an introduction', in *Securing Access to Health Care. Volume Two: Appendices. Sociocultural and Philosophical Studies*, U.S. Printing Office, Washington, D.C.

Chapter 13

Justice, health care, and the elderly

A dominant theme in both public and policy-making discussions of health care in recent years has been the need to control relentlessly escalating health care costs. One widely perceived and important cause of this problem has been the "graying" of America – recent and anticipated increases in life expectancy and in the proportion of the population over age 65, and especially the even more rapid increase in the "old old" population. The increased numbers of elderly in turn make disproportionately large use of health care. In response to any proposals to limit the availability to the elderly of resources generally, and health care resources in particular, their advocates have added the charge of "ageism" to the more familiar charges of racism and sexism. From the other side and in the face of the daunting political power of the elderly wielded by such lobbying groups as the American Association of Retired Persons, the argument is increasingly heard that making Social Security or Medicare programs relatively immune from the budget cuts suffered by other domestic programs is creating intergenerational injustice – that the welfare of children and working-age Americans is

A review of Norman Daniels, *Am I My Parents' Keeper?* (Oxford and New York: Oxford University Press, 1988), and Daniel Callahan, *Setting Limits* (New York: Simon and Schuster, 1987). References to these works appear as (D) and (C) respectively. An earlier version of this review essay was prepared for a symposium at the American Philosophical Association Pacific Division meeting in 1988. I am grateful to Daniels, Callahan, and the Editors of *Philosophy & Public Affairs* for helpful comments.

being wrongly sacrificed for the benefit of the elderly. The intensifying policy debate over health care and other social support programs serving the elderly is forcing a rethinking of questions of intergenerational justice and the claims of the elderly on social resources.

Norman Daniels and Daniel Callahan have written books that address these issues. While the arguments of both, and especially of Daniels, extend beyond health care, their central focus is on what I will here call the age-group problem – what ethical claims do the elderly, as compared with other age groups, have on social resources for the provision of health care? Both conclude that some direct appeals to age in the determination of these claims can be ethically unobjectionable, though the nature of their arguments as well as the conclusions they reach are in many important respects very different. Daniels's argument is squarely within the tradition of political liberalism, while Callahan calls on a communitarian political philosophy. Despite these differences, I will argue that central parts of what is most interesting and plausible in each's view are compatible and that not losing sight of them will yield a richer and deeper public debate about health care for the elderly. Precisely because of these differences in broader political philosophy, my disagreement with Callahan runs considerably deeper than my disagreement with Daniels, and I shall criticize some of the features of Callahan's position that have generated the most controversy. I believe it would be unfortunate, however, if public attention and policy discussion focused only on some of his more controversial and problematic ethical claims and policy proposals, thereby losing sight of what I consider an important contribution to this debate. In Daniels's case, I have less to disagree with, but will show why the nature of his overall theory of justice in health care leaves him largely silent on the important question Callahan addresses.

One possible ground of obligations to meet the needs of the elderly is filial obligations, but both Daniels and Callahan reject this route. Whatever the truth about how much such care was provided in the past either in our own or in other societies, tradition alone cannot guide us now because of the great increase in the needs of the elderly – a greater proportion of the population lives into old age, those who do so live a longer period of old age, and, because medical care can and often does sustain their lives during long periods of disability and dependence, their needs for care while elderly are commonly much greater than in the past. Yet it is precisely in cases where the needs of the elderly are so great as to overwhelm the capacities of their grown children to pursue their own lives while meeting those needs that an account of how much care is owed the elderly and by whom is most needed. Both Daniels (D, pp. 28–34) and Callahan (C, pp. 84–106) argue that no general moral argument succeeds in showing that children are obligated to provide the care their parents need. Moreover, there is no consensus among actual families on the care and aid that children owe their elderly parents which might instead guide public policy.

It is worth adding that even if there were a single sound argument establishing what aid and care children owe their parents, this could not by itself adequately guide public policy concerning care for the elderly, for the simple reason that it would not tell us what care is owed the elderly who are childless, or whose children are no longer living or are otherwise unable to provide the care. Nor could such an argument ground a social policy supporting the childless elderly in the manner and to the degree that elderly parents are owed support by their adult children, since it would presumably be the actual care given children that would obligate the children in turn to care for their elderly parents. The elderly who have never had children have never provided any care on which a reciprocal obligation to care for them could be based.

How then are the claims of the elderly to socially supported health care to be determined? I believe the conventional view in medicine is that there is no *special* problem of what health care justice requires be made available to the elderly as opposed to other groups. There is only the general problem of determining the adequate level of health care that justice requires be available to *all* citizens. Both physicians and health policy analysts commonly urge that only patient *need* should be relevant to allocating health care; in conditions of scarcity we must simply meet the most important or urgent health care needs. In this view, the health care needs of the elderly, like those of all others, should be considered solely in terms of their importance or urgency and not discounted simply because they are the needs of the elderly. Of course, both the health care needs and the expected benefits of treatment of many elderly may vary in systematic ways from those of younger patients, but it is these differences, not differences in age, that are relevant for determining their claims to social resources for treatment. In the conventional view, differential treatment based on age itself is unjust ageism. Both Daniels and Callahan reject this conventional view, though for very different reasons.

In an earlier book Daniels developed a general theory of just health care within a broader liberal political philosophy of a generally Rawlsian form.[1] Now he extends that general theory to the problem of what health care is owed the elderly as opposed to other age groups. His account exploits the central fact that we age and so in the course of our lives can expect to pass through each of the stages or age periods of a normal lifespan. As a result, the much-discussed competition for resources between young and old is misleading in suggesting a conflict between different groups of persons. Instead, it is possible to frame the question of distribution of resources between different age groups as a problem of prudential reasoning for a single individual about how to distribute resources over the different stages of his or her own life. Daniels calls this the "prudential lifespan account." Institutions such as Social Security and health care insurance plans take more resources from workers in their middle, productive years than are returned to them then in benefits, but in turn provide more in benefits during retirement years than are then taken in taxes or premiums. Instead of thinking of these as *inter*personal group transfers, we can regard them as *intra*personal savings plans for persons as they pass through these institutions over the course of a lifetime (D, pp. 42–47). At any particular stage of life, all persons will be treated the same, though different birth cohorts will reach a particular stage of life at different times.

The prudential lifespan treatment of the age-group problem presupposes that we have both a general theory of distributive justice and a theory of just health care, which together determine what is a fair lifetime share of health care for members of a particular society at a particular time. Specifically, Daniels assumes, first, that the general theory of justice will include some principle of fair equality of opportunity requiring that "opportunity be equal for persons with similar skills and talents" (D, p. 70). Second, he assumes that the importance of health care for justice is determined by its effects on a person's opportunity, and so a fair share of health care is determined by what is necessary to maintain fair equality of opportunity over a lifetime. Prudential planning for health care needs will then require an individual to choose a particular set of institutions that will distribute health care to her as she lives her life and passes through these institutions. Prudence requires neutrality towards all the stages of one's life, and not biasing the choice of institutions to favor one's present stage. For this, a Rawlsian veil of ignorance regarding one's present age or life stage would seem a natural device. We might think of these prudential planners as knowing their aims and values or plans of life at all stages of their lives, while the veil of ignorance only deprives them of knowing what stage is "now." However, because our plans of life change over the course of our lives, very often in major and largely unpredictable ways, planning based on knowledge of what our life plans will be over the course of our lives is not possible.

Instead, Daniels argues that the neutrality that prudence requires towards our different life stages can be obtained by seeking to provide ourselves with a roughly equal opportunity at each life stage to revise and carry out our plans of life then, whatever they turn out to be. Prudential planners can achieve this by securing for themselves a bundle of Rawlsian primary goods that will provide them with a fair share of the normal opportunity range for their society at each stage of life.[2] In then planning for health care at different life stages, it will be rational for them to take account of the general prevalence of different diseases at different stages of life, of the effects of those diseases on the range of normal opportunity at particular stages of life, and of the possibilities and costs of treating those diseases. For example, Daniels shows how the prudential planners' focus on opportunity supports a substantial strengthening and expansion of long-term care services for the elderly (D, chap. 7). Prudential planners can be expected to provide for significantly different kinds and amounts of health care at different stages of life, since the nature and prevalence of disease, the effects of diseases on normal opportunity, and the benefits and costs of treating disease vary substantially at different life stages. Because their lifetime fair share of health care will not provide all possible beneficial care for all their health care needs without regard to cost, all of these age-related differences will affect how they allocate between different life stages in these conditions of moderate scarcity.

Since prudential planners allocate their lifetime fair shares of health care relative to needs at different life stages, it may not be clear how Daniels's prudential lifespan approach differs from the conventional appeal to needs. The difference can be brought out by considering what Daniels calls "pure age rationing" – making differential treatment available to persons of different ages not because they have different health care needs, but simply and only because they are members of different age groups. The most important and controversial case is life-sustaining treatment.

Daniels argues that his prudential planners could face choices under conditions of substantial scarcity in which an increased access to life-extending resources in one's later years requires a reduced access to them in earlier years; greater use of these resources late in life will require increased saving in one's productive years. For example, increased access to acute care interventions employing expensive, high-technology care, such as treatment in an intensive care unit, could require cuts in prenatal care leading to an increase in infant mortality rates or cuts in immunization programs for potentially fatal infectious diseases that strike early in life.[3] His example compares scheme A (for age rationing), in which no one over age 75 is offered any of several high-cost life-extending technologies (such as dialysis, transplant surgery, or extensive bypass surgery), and scheme L, in which allocation is strictly by medical need, so that only one of the high-cost technologies can be developed and made available to all who need it (D, pp. 87–88). To simplify, assume that A offers a 1.0 probability of living to age 75 and then dying immediately, while L offers a 0.5 probability of living only to age 50 and a 0.5 probability of reaching age 100 and then dying. Life expectancy is identical in A and L, with the difference that L trades some expected life years before one has reached the normal lifespan of 75 for the possibility of life years beyond the normal lifespan. Scheme A differs from the conventional appeal to health care needs because under A a person below age 75 will receive an expensive treatment like bypass surgery, while someone over 75 will not get it, though both may have the same coronary disease and medical need for and expected life extension from the surgery.

Daniels offers two reasons why prudent planners would prefer A – the age-rationing scheme (D, pp. 89–91). First, they would know that the incidence of disease and disability is greater between ages 75 and 100 than between 50 and 75, and so the *quality-adjusted* expected life years are commonly greater under A than under L. Second, they would know that while in some lives and life plans the "golden years" beyond age 75 may even be among a person's best years, in most life plans important plans and projects will be largely completed by age 75, making the 50-to-75 period more important than the 75-to-100 period. This shows that prudent planners could prefer A – the age-rationing scheme – and that age rationing is not per se unjust in comparison with allocation according to medical need. The one paragraph

Daniels devotes to the second reason, however, quite un-
derplays its importance (D, pp. 90–91).

Callahan's notion of a "tolerable death" is helpful in bringing out that importance. A tolerable death is a "death at that
stage in a lifespan when (a) one's life possibilities have on
the whole been accomplished; (b) one's moral obligations to
those for whom one has responsibility have been discharged;
and (c) one's death will not seem to others an offense to
sensibility, or tempt others to despair or rage at the finitude
of human existence" (C, p. 66), and, he adds, (d) the death
must not be "marked by unbearable and degrading pain"
(C, p. 72). Such a death "may be understood as a sad, but
nonetheless relatively acceptable event" (C, p. 66). This is,
I believe, an important notion which seeks to distinguish, at
least with conditions (a)–(c), deaths that we accept as simply
a normal and inevitable part of the human condition – the
mortality we all share – from those that we tend to characterize as a tragic or unfair cutting short of a life in process.
As Callahan notes, this notion of a tolerable death appeals
to the idea of a life in a narrative or biographical, not simply
a biological, sense (C, p. 66). A tolerable death occurs at a
point at which what a person can largely expect from and
within a human life is not denied by the death, though it is
still a sad event and a loss for the person whose death it is
and for the others who cared about the person. Our own
death is the central and overriding inevitability in our lives
(more inevitable even than taxes – a few do succeed in evad-
ing them). However consciously and explicitly, our lives, and
the plans we make for them, are fundamentally shaped by
what we can expect in the way of a normal lifespan. While
death can be hard and may be resisted at any age, it is a
death before one has had the opportunity to complete a normal lifespan that will often seem an unexpected and bitter
deprivation.

Callahan's notion of a tolerable death underscores the importance of the second reason Daniels offers for why an age-
rationing scheme need not be unjust – we plan our lives and

what will fill them based on expectations of what is a roughly normal lifespan in our society. Because we commonly try to give our lives continuity and coherence, as biographical lives lived by us from the "inside," the loss of life years needed to reach a normal lifespan and "complete that biography" is commonly a significantly greater loss than the loss of a comparable number of life years needed to live beyond the normal lifespan. Thus, if we simply look to the good we can do for persons whose lives are extended by the provision of life-sustaining medical care or by other health policies, we can often do substantially more good by preventing deaths before instead of after the normal lifespan has been reached.

It may be objected that this claim is class-biased, and true only for persons whose work is satisfying and rewarding, but not for the many whose work is largely drudgery and who may look forward to relief from it in retirement. Two points largely, though not completely, undercut the force of this objection. First, a normal lifespan of at least 75 to 80 years already includes a significant number of these retirement "golden years," including those most likely to be free of disability. Second, even for many whose work may be largely drudgery, other central and satisfying projects of their lives, such as raising their children, are completed within the normal lifespan.

The distinction between a premature and a tolerable death suggests that an argument based on a moral principle of fair equality of opportunity could directly support assuring the life years necessary to live out a normal life, at least where doing so is technologically and socially possible. Such an argument would be independent of whether the prudential lifespan account is accepted for the age-group problem. This opportunity-based moral claim would not be absolute, in the sense of being unlimited no matter what the cost of meeting it, just as Daniels's broader account of a person's just claims to health care grounded in fair equality of opportunity is not unlimited no matter what the cost of meeting it. My suggestion is that the notion of fair equality of opportunity might be extended not just to apply to health care generally, as

Daniels did in *Just Health Care,* but also to morally differentiate the claims to life-extending health care up to, as opposed to beyond, the normal lifespan. Daniels rejects this line of reasoning (D, pp. 92–93). But it is, I believe, a plausible use of the distinction between a premature and a tolerable death, which should fit within broader theories of distributive justice that include a principle of fair equality of opportunity. It should not conflict in major ways with the substantive implications of Daniels's prudential lifespan account, but instead has some promise of providing an independent moral argument that would largely reinforce at least some of the broad implications of his account. Both arguments establish that under some conditions of scarcity what I will call *weak age rationing* – giving greater weight to the claims on social resources for life-sustaining care up to, as opposed to beyond, the normal lifespan – need not always be unjust.

In this respect, more use can be made than Daniels does of a principle of fair equality of opportunity for determining a just allocation of resources to life-extending health care. In a different respect, however, I believe Daniels's theory places too much moral weight on opportunity in a way that his own prudential lifespan account highlights. His treatment of the age-group problem here inherits a difficulty in his more general theory of just health care. Some critics of that theory have argued that it is a mistake to hold that the relevance of health care to justice lies only in its effect on opportunity.[4] Instead, health care is important also for its effects in relieving or preventing pain and suffering and in improving the quality of life in respects independent of opportunity. Prudential planners would allocate some of their fair share of health care resources to achieve these non-opportunity-based ends. But if such ends have been ignored in the determination of individuals' fair lifetime shares of health care resources, then there is an incoherence between Daniels's general theory of just health care and his prudential lifespan treatment of the age-group problem. Prudential planners will want to allocate their fair lifetime shares of health care resources in part to serve ends other than opportunity for which no provision

397

was made in the opportunity-based determination of those shares. This difficulty, however, lies in Daniels's general account of just health care, and with appropriate revision of that account the prudential lifespan treatment of the age-group problem is, in my view, essentially sound.

While Daniels's treatment of the age-group problem is squarely within the tradition of broadly individualist and pluralist liberalism, as represented within political philosophy most prominently by Rawls, Callahan rejects central features of that tradition. Daniels's theory of just age-group distribution, like Rawls's broader theory of justice, rejects appeal to any one substantive account of the good life, appealing instead to the liberal view of justice as establishing the terms of social cooperation within which individuals are left free to choose and pursue their own inevitably differing plans of life. Thus, in Daniels's theory persons will seek to secure a bundle of primary goods, including health care, that will provide them with roughly equal opportunity at all stages of their lives to pursue their conception of the good life, whatever it happens to be and as it changes over their lives. Against this aspect of traditional liberalism, Callahan holds that social policy towards the elderly should be based on a communal consensus on the proper meaning and ends of aging and old age.

Callahan is especially concerned to reject what he calls the modernized view of old age, which sees it as providing simply more of the same kind of life, emphasizing the pursuit of private aims, as when one was young (C, pp. 26–31). The modernized view seeks essentially to banish old age and to avoid death as long as possible, if not to deny death's existence altogether. It denies the limits that aging and death place on our lives, to which Callahan wants to give due place. The modernized view of old age, he argues, will inevitably fail because however much the wonders of modern medicine may delay the onset of decline in our powers and our ultimate death, it cannot succeed in banishing them. We need a different view of the meaning and ends of old age that acknowledges it as the period when our lives, as seen both in the biological sense and from the inside in a biographical sense, are drawing to a close and reaching a completion.

According to Callahan, the primary purpose of old age and of the old should be the service of the young, of the generations that will follow (C, p. 43). By virtue of being near the end of their own lives, lives which can be seen as integrated and coherent wholes, the elderly are uniquely situated to integrate past with present, to understand the way past, present, and future interact, to show what it means to live in the present and not always for the future, and to help the young come to understand what it is to live a meaningful life that does not need continually to deny that all such lives must inevitably come to an end (C, pp. 44–48). For the elderly to play this role, Callahan repeatedly emphasizes, a fundamental and deep change will be required in our society's common view of the place of aging in a complete life, a change that if it comes at all will take at least a generation of public debate to bring about (C, pp. 199–200). Changes will be required both in the way the young view aging and the old and in how the elderly themselves view their current stage of life. Callahan believes that such a transformation in the way aging and old age are viewed would benefit the elderly in giving them a richer, more satisfying view of the meaning and ends of aging and of a complete life.

What would follow from this transformed view of aging and old age for health care policy, specifically as it concerns allocation of resources for the elderly? In the broadest terms, Callahan argues that the suitable goal for medicine would be the achievement by all of a natural lifespan and, beyond that, only the relief of suffering. Government would have no obligation to help people extend their lives medically beyond the natural lifespan, no responsibility to provide "any medical intervention, technology, procedure, or medication whose ordinary effect is to forestall the moment of death, whether or not the treatment affects the underlying life-threatening disease or biological process" (C, pp. 137–38). At the same time Callahan argues, like Daniels, for substantial expansion in the provision of long-term care and support services for the elderly. Thus, it is a mistake to view this change in perspective as merely a thin guise for cutting spending on the elderly – while it would support cuts in some areas, it would require major new commitments of resources in other areas.

Callahan in fact displays considerable ambivalence about the relation of scarcity, and an increased awareness of scarcity, to his proposal. On the one hand, he notes the familiar data on the expanding numbers and proportion of the elderly in society, as well as their disproportionate and rapidly growing use of health care resources, and comments more than once that we cannot continue to pour resources into extending life and postponing death (C, pp. 21, 137; appendix). On the other hand, he also repeatedly notes that his motivation is *not* simply to reduce health care expenditures, but instead to provide a revised view of aging and old age that will be a benefit to and improve the lives of the elderly as much as the young (C, pp. 53, 116).[5] Moreover, he explicitly adds that it would be unjust to restrict provision of care to the elderly merely in order to save money and in the absence of this change in the view of aging and the elderly (C, p. 153).

Callahan suggests several implications of this revised view of aging and old age for how we set priorities in public policy towards the elderly. First, "no new [medical] technology should be developed or applied to the elderly that does not promise great and inexpensive improvement in the quality of their lives, no matter how promising for life extension" (C, p. 143). Second, the system of providing equitable security to the elderly against the loss of financial and other independence must be greatly strengthened (C, pp. 146–48). Third, priorities for research and care should be given to the causes of premature death, chronic diseases that burden the later years, and support services that reduce suffering and increase the quality of life, such as improved home care (C, pp. 148–53).

Much of this reorientation is well taken, but clearly most controversial in Callahan's proposal is his support of *strong age rationing* – the elderly should have *no* entitlement to social resources for the provision of life-extending care once they have reached a normal lifespan, that is, the late 70s or early 80s. (Callahan explicitly does not support what can be called *very strong age rationing*, which would prohibit persons even from using their own resources to purchase such care. He explicitly allows for private markets in this care, but notes the moral and political difficulties facing such a two-class system.) Precisely how is this strong age rationing of life-extending treatment beyond the normal lifespan to be justified? I believe Callahan's answer is related to his dissatisfaction with pluralistic liberalism. Against that liberalism, he insists that it is necessary to reach a *social* consensus on the meaning and ends of aging and old age, a consensus on a *single* view of their meaning and ends (C, p. 220). He insists on this because he believes that we require a social policy concerning health care for the elderly, and that such a policy's legitimacy must lie in a social consensus it both rests on and expresses. If a "consensus" is understood along very strict lines as meaning that essentially *all* members of society share the view, and the consensus is informed and has been freely arrived at by those members, it would then not be necessary to impose the policy of not providing life-extending care to persons beyond the normal lifespan on anyone who does not accept that policy, even if weakness of will may lead some to seek to get such care when they come to need it. If all freely and knowingly accept the policy, then it is unnecessary or superfluous to provide any further argument that justice requires it. But is there any reason to expect such a strong consensus to come about in this country? I think the answer must clearly be no.

People now find meaning in their later years in many different ways and pursue a host of different ends in those years. Moreover, that seems inevitable in a large and diverse country that celebrates its pluralism and gives such high value to individual liberty. Since a strong consensus on the meaning and ends of aging and old age would likely require deep and pervasive social transformations, including deep restrictions on highly valued liberties and on individual self-determination in shaping one's own conception of the good, we have good reason to resist steps for achieving such a consensus in the United States. But in the absence of such a consensus, we need a justification for the denial of all social entitlements to life-extending care beyond the normal life-span for those who reject the consensus and want that care. Since Callahan's only justification seems to suppose this missing consensus, does that mean his position must be systematically rejected? I think not.

Callahan's book has forced into the foreground of the debate about health care and the elderly the extremely important questions that *individuals* must face about how they will find meaning in old age and what ends will guide that stage of their lives. Many will no doubt reject the answers that he offers to these questions. But many others will find his criticisms of the modernizing view important and the alternative ideal of aging and old age that he offers deeply attractive. That his answers to these questions will achieve no full consensus does not detract from the importance of the questions or the answers. In fact, I believe it is here that his book's greatest strength and importance lies. Nevertheless, the absence of any strong social consensus along these lines undermines Callahan's appeal to a communitarian basis for at least the restrictive aspects of his policy proposals regarding social entitlements to health care for the elderly, and specifically to life-extending care beyond the normal lifespan. In seeking a basis for public policy in a pluralistic society like ours that lacks, and will continue to lack, any strong social consensus on the proper meaning and ends of aging and old age, I believe we must turn to liberalism and an approach largely along the lines of Daniels's prudential lifespan account. Specifically, it is to that approach that we must turn in thinking about just distributions of resources between different age groups. However, just because liberalism appeals only to a thin theory of the good for persons and rejects any appeal to a single, thick account of the meaning and ends of life in general, or old age in particular, we will find it largely silent on these issues of most importance in Callahan's book; Daniels's book in particular says little on them.

It is then in the following respect that Daniels's general theoretical framework – the prudential lifespan account – for just age-group distribution is compatible with Callahan's proposals concerning the meaning and ends of old age. Those proposals offer what many persons living within an institutional framework whose age-group distributions meet the requirements of the prudential lifespan account would find an attractive and appealing view of the meaning and ends of old age for themselves and their own lives. For carrying out this view that they have freely adopted of their old age in the context of specific decisions about their medical treatment, should they later be unable to decide for themselves, such persons might employ advance directives such as living wills and durable powers of attorney for health care. Some of Callahan's other policy recommendations and priorities might be revised so as to affect only those who have freely adopted his view of the meaning and ends of old age, though I shall not explore that possibility further here. But this is not to say that these restrictions on social entitlements to life-extending care should be imposed as a matter of public policy on persons who have not freely adopted this view of old age as their own. For determining just age-group distributions, Daniels's prudential lifespan approach is the more reliable guide.

Before concluding, I want to say a bit more about the specific policy issue of life-extending care beyond the normal lifespan. This issue has understandably been the focus of a great deal of attention, though other proposals like improving long-term care and home care services for the elderly, on which, as I have noted, Daniels and Callahan agree, would probably have a greater overall impact on the well-being of the elderly. Can we use prudential lifespan reasoning to bring out what is wrong with Callahan's proposal to deny *any* social entitlements to life-extending care for patients beyond the normal lifespan? A prudential planner under a suitable veil of ignorance, deciding whether to make provision or insure for life-extending care beyond the normal lifespan, would consider at least the following: What is the relative frequency of various life-threatening conditions in persons beyond the normal lifespan? What is the cost of various life-extending treatments for such conditions, and what other goods would have to be generally forgone if those costs are to be borne? What is the expected length and quality of life extension from such treatments, and what is the relative importance of the typical plans and projects that could be pursued or completed during that period of life extension?

It is obvious that a complete analysis would require a great deal of information, much of which we lack in any reliable form. But it is not hard to see that there will be many possible (and actual) cases of life-extending care that a prudential planner would provide for under the conditions of moderate scarcity that obtain in the United States today. To take an extreme example, consider a very healthy and vigorous 82-year-old writer who continues to be actively engaged in several writing projects, has a full and satisfying family life, pursues a number of important community activities for the benefit of others, and greatly enjoys periodic travels. She develops a pneumonia that is life-threatening without treatment, but that can be relatively simply and inexpensively treated with a short hospitalization and course of antibiotics. There is every reason to believe that with prompt treatment there will be no significant long-term impact on her health and that she will be able to return quickly and without any significant deficit or disability to her previous mode of life. Given her otherwise excellent state of health, with treatment she can expect to live another decade.[6]

Can there be any doubt that a prudential planner would devote the limited resources necessary to secure care in circumstances such as these? If not, then it is equally clear that no *general* denial of social entitlements to life-extending care for persons beyond the normal lifespan can be derived from the prudential lifespan account of just age-group distribution. And while this case is admittedly extreme in the very substantial benefits promised by treatment at a minimal cost, I believe many other, more common cases would also promise sufficient benefits, given their costs, for a prudential planner to provide for the treatment. (It is unclear why very high benefit/low cost health care is substantially different from other high benefit/low cost necessities of life for a healthy person, such as food. It is therefore unclear why Callahan's reasoning in support of denying to persons beyond the normal lifespan all entitlements to social resources for life-extending *health* care would not apply to the use of social resources for other life extension needs such as *food* as well, though he certainly would not welcome this implication.) Each of the factors that would determine choice varies along a broad continuum, but the closer they get to the area in which a prudent planner would likely judge the benefits of life-extending care not worth the resources it requires, the closer one also is to the cases in which actual persons able to decide about treatment, or to give advance directives about their wishes, would not want the treatment for themselves regardless of whether resources were available to pay for it.

Thus, at current levels of overall resource scarcity in this country, it is not clear that prudent planners with a lifetime fair share of health care resources would exclude coverage for substantial amounts of life-extending health care beyond the normal lifespan. Their attention would better focus, first, on developing and utilizing means of maintaining control over what life-sustaining treatment they receive, especially should they become unable to make decisions about their treatment, and, second, on programs that improve the quality of lives of the elderly. However, while I am skeptical about cost-containment measures narrowly focused on life-sustaining treatment for the elderly, nothing I have said should be taken to indicate opposition to other, differently focused cost-containment measures. These must be evaluated in their own terms, and I believe many are ethically justified. But the shameful inadequacies in many of our social welfare programs, including those designed to ensure that all citizens have basic goods such as food and shelter, as well

as health care, are results of failures of political will, not of
our failure to adopt strong age rationing of life-sustaining
health care.

NOTES

1 Norman Daniels, *Just Health Care* (Cambridge: Cambridge University Press, 1985).
2 Daniels defines the general notion of "the normal opportunity range for a given society" as "the array of life plans reasonable persons in it are likely to construct for themselves" (D, p. 69). The prudential planners disaggregate this notion and employ an *age-relative* opportunity range for different life stages.
3 It should be added that Daniels appears to believe that scarcity now is not severe enough to force such choices. Whether this is correct will depend on complex questions concerning trade-offs both between different health care resource uses and between health care and other resource uses. In the context of ideal theory, it may be defensible to assume no restrictions on possible resource trade-offs available to planners. However, for real-world policy planners the level of scarcity assumed will depend on an assessment of feasible policy alternatives given political and other constraints on possible reallocations. Here it will not do to resist claims of scarcity of health care resources on the grounds, for example, that "Star Wars" weapons funds would be better spent on health care, unless such a reallocation is feasible for policy-makers.
4 For example, Allen E. Buchanan, "The Right to a Decent Minimum of Health Care," *Philosophy & Public Affairs* 13, no. 1 (Winter 1984): 63; and L. Stern, "Opportunity and Health Care: Criticisms and Suggestions," *Journal of Medicine and Philosophy* 8 (1983): 336–62.
5 In this respect, it is worth noting that Callahan was defending the main features of his view before the current obsession with cost containment developed. See his "On Defining a 'Natural Death,'" *Hastings Center Report* 7 (1977): 32–37.
6 Callahan makes an exception to his proposal for strong age rationing of life-extending treatment for just such a case as this – what he calls "the physically vigorous elderly person" (C,

p. 184). But he offers no principled basis for this exception to his general view, noting only that he does "not think anyone would find it tolerable to allow the healthy person to be denied lifesaving care" (C, p. 184).

Chapter 14

Truth or consequences: The role of philosophers in policy-making

My reflections here are based principally on my experience during the 1981–82 academic year as staff philosopher on the President's Commission for the Study of Ethical Problems in Medicine, as well as participation in various capacities in other policy-making or advising bodies in biomedical ethics at both state and national levels. My central thesis is that there is a deep conflict between the goals and constraints of the public policy process and the aims of academic scholarly activity in general and philosophical activity in particular. I shall support this thesis by developing several related aspects of the conflict.

Truth is the central virtue of scholarly work. Scholars are taught to follow arguments and evidence where they lead without regard for the social consequences of doing so. Whether the results are unpopular or in conflict with conventional or authoritative views, determining the truth to the best of one's abilities is the goal. In philosophy, especially, nothing is to be immune from question and criticism; all assumptions are open to and must withstand critical scrutiny. Now it would be silly to maintain that philosophers always succeed in this unconstrained quest for the truth, either in the sense that their quest is unconstrained or that they reach

I want to acknowledge here my very substantial indebtedness to Daniel Wikler, whose talk to a Brown University bioethics conference first set me to thinking about philosophers and policy-making. We have had many subsequent discussions of these matters, and I am sure many of the ideas here probably originated with him, though I am no longer sure which.

the goal of the truth. We often fail to recognize the problematic nature of particular assumptions or views and so fail to subject them to the criticism they deserve. Like our colleagues in the natural and social sciences, we can become wedded to particular views or general theories so that we fail to recognize or acknowledge the difficulties facing them. At any time, much is simply beyond the grasp of our best efforts.

When philosophers become more or less direct participants in the policy-making process and so are no longer academics just hoping that an occasional policymaker might read their scholarly journal articles, this scholarly virtue of the unconstrained search for the truth – all assumptions open to question and follow the arguments wherever they lead – comes under a variety of related pressures. What arises is an intellectual variant of the political problem of "dirty hands" that those who hold political power often face. I emphasize that I do not conceive of the problem as one of pure, untainted philosophers being corrupted by the dirty business of politics. My point is rather that the different goals of academic scholarship and public policy call in turn for different virtues and behavior in their practitioners. Philosophers who steadfastly maintain their academic ways in the public policy setting are *not* to be admired as islands of integrity in a sea of messy political compromise and corruption. Instead, I believe that if philosophers maintain the academic virtues there they will not only find themselves often ineffective but will as well often fail in their responsibilities and act wrongly. Why is this so?

The central point of conflict is that the first concern of those responsible for public policy is, and ought to be, the consequences of their actions for public policy and the persons that those policies affect. This is not to say that they should not be concerned with the moral evaluation of those consequences – they should; nor that they must be moral consequentialists in the evaluation of the policy, and in turn human, consequences of their actions – whether some form of consequentialism is an adequate moral theory is another matter. But it is to say that persons who directly participate in the formation of public policy would be irresponsible if they did not focus their concern on how their actions will affect policy and how that policy will in turn affect people.

The virtues of academic research and scholarship that consist in an unconstrained search for truth, whatever the consequences, reflect not only the different goals of scholarly work but also the fact that the effects of the scholarly endeavor on the public are less direct, and are mediated more by other institutions and events, than are those of the public policy process. It is in part the very impotence in terms of major, direct effects on people's lives of most academic scholarship that makes it morally acceptable not to worry much about the social consequences of that scholarship. When philosophers move into the policy domain, they must shift their primary commitment from knowledge and truth to the policy consequences of what they do. And if they are not prepared to do this, why did they enter the policy domain? What are they doing there?

Let me be more specific about some of the forms this conflict between scholarly and policy goals and virtues has taken in my own experience. No philosopher that I am aware of has ever been an omnipotent philosopher king or queen – able to make public policy however he or she sees fit. Instead, I think my own experience is more typical. I worked with nonphilosophers on the professional staff of a presidential commission, and the staff worked for the commissioners who were political appointees out of the office of the president. We published ten book-length reports on different issues in biomedical ethics, each of which had several staff members assigned to it. The Commissioners had the final word on what our reports would say. Thus, though I had my own views about what they should say, those views would only have any effect if I was able to persuade other staff members working on a particular project, and in turn the Commissioners, of them. I and the two other staff philosophers who preceded and followed me were, quite rightly, not accorded the role of moral authorities to whom appeals were made for the right answers; our impact lay instead in our ability to persuade.

The staff was also often more expert than our "bosses," the Commissioners, on many of the problems on which we worked. In our case, as elsewhere in the policy and political world, this was to a large extent inevitable and appropriate since the staff was selected for their professional expertise in the area of the Commission's responsibilities and worked full time on our projects, while the Commissioners retained their other usual professional responsibilities and spent only a small portion of their time on Commission work. Staff sometimes believed that Commissioners held particular views on indefensible grounds. If our reports were to say what we thought they should say, we had to bring the Commissioners around to our views. It was in the resulting context of debate and dialogue that I and other staff members often found ourselves looking to what the consequences on others would be of making a particular argument or taking a particular position, instead of simply at whether we considered the argument or position sound. The goal often became to persuade or even to manipulate others in order to reach a desired outcome instead of a common search for the truth. There is space to give only one example.

In our report on decisions about life-sustaining treatment, we addressed briefly a number of distinctions that commonly play a role in the reasoning underlying those decisions, distinctions such as between killing and allowing to die, between a physician's or a disease's being the cause of death, and so forth. I believe that on common understandings of the kill/allow to die distinction, the difference is not in itself morally important, and that stopping life-sustaining treatment is often killing, though justified killing. Needless to say, many of the Commissioners did not share this view. They believed that killing was far more seriously wrong than allowing to die, and that stopping life support was allowing the patient to die of his disease, not causing his death and killing. We shared the conclusion that stopping life-sustaining treatment at the request of a competent patient was morally permissible, but I believed that their reasons for this conclusion were confused and unsound and that I might have some success in convincing them of this. My philosophical instincts urged me to attack the confusion and to follow the argument wherever it led.

But what would be the consequences of convincing them either that allowing to die is in itself no different morally than killing and/or that stopping life support was killing? A quite plausible case could be, and was, made that this could throw into question their acceptance of the moral permissibility of stopping life support. Could one then responsibly attack what seemed confusions in their view when the result of doing so might well be to lead them to an unwarranted and worse conclusion – and a conclusion, it is important to add, that could produce important adverse consequences in suffering and loss of self-determination for real people? I want to stress that this attention to the consequences of criticizing a position or defending one in a particular way became a significant factor in our work. The example I have cited was not an isolated case but only one instance of a common phenomenon. I believe that when a philosopher or anyone else accepts the role and responsibilities of participating directly in the policy-making process, it would be morally wrong and irresponsible not to give substantial weight in this way to the consequences likely to flow from one's actions in that role. Doing so, however, leads to manipulative attitudes toward others that I am not comfortable with and fosters playing a little fast and loose with the truth as best one understands it, in a way that is inimical to the scholarly academic enterprise.

There are related aspects of this general conflict. Philosophers are viewed as somewhat strange beasts in governmental and policy circles. It is never clear to many others exactly what they do or what are the criteria for their having done it well. There is particular skepticism about whether academics in general, and especially philosophers, understand how the "real" political world works and the constraints it puts on policy-making and policy. All of this means that philosophers have a credibility problem in policy circles that leads to pressure for some cutting and trimming in at least the voicing of one's more extreme, unconventional, or radical views. To voice and press views that others will find outrageous, however much one may be convinced of them, is to risk using up one's credibility and to risk not being heard, or even losing the opportunity to speak, on other occasions when one might have a significant impact. Once again, the effect is to make philosophers look over their shoulders at what effect pressing their views will have on others, independent of whether they believe them to be sound.

It is not just the inherent conservativeness of policymakers and bureaucrats that creates pressure not to voice views that others will find extreme or bizarre. An important part of the policymaker's job is to "sell" a position or policy to others in the policy and political process. Ultimately, it must be sold to all others affected by and affecting the process, including the public. That makes the "packaging" of a policy proposal often extremely important to its fate. The particular formulation and defense of a policy that is likely to move it most successfully through the policy arena may differ substantially from what a philosopher believes its correct formulation and defense to be. For example, philosophers who believe that infanticide is morally permissible would be ill advised to use that as the basis of an attack in the public policy arena on the Reagan administration's so-called Baby Doe regulations.

On this point, as elsewhere for the general scholarly-policy conflict I am sketching, the contrast should not be overstated. Philosophers in their academic scholarship are not indifferent to whether they convince others of their views. But the attempt of scholars to persuade others is not just a process of coalition building to gain allies. What is important is that other scholars are persuaded by their own assessment of one's arguments and evidence.

A related aspect of moral philosophers' credibility problem lies in their own uncertain views of the nature of their enterprise, as much as others' views of it. For many philosophers, basic moral principles cannot ultimately be established as true, objectively correct, and so forth. Rather, the deepest moral conflicts and disagreements may admit of no rational resolution. But if so, then philosophers' grounds for using their expertise or title to press their own views on moral questions may in turn be uncertain. Yet they inevitably will do so, even if neither they nor others construe their role to be provider of the moral truth and of solutions to moral problems. One can hold, as I do, some form of coherentist view of justification in ethics without slipping into a radical moral skepticism or subjectivism, but for many nonphilosophers coherentism never seems a fully convincing or adequate account of moral justification.

A further aspect of the scholarly-policy contrast I have been sketching is what Norman Daniels has called the priorities problem, and what I will call the agenda problem.[1] The problem is what is to be taken as fixed or given for the purpose of setting or changing policy and what is to be taken as open to modification and so on the policy agenda. The scholarly philosophical virtues I have sketched above are intended to leave nothing fixed or given, and beyond criticism, revision, or rejection. This leads the philosopher toward a maximally wide agenda; no change is too far-reaching if persuasive argument supports it. For seasoned policymakers and bureaucrats, on the other hand, who have lost as many battles as they have won and who are constantly subject to the competing forces and interest groups active in the policy process, all is not immediately possible. Many issues are not on their agenda because they are not politically feasible, or because it is not an opportune time for them, or because efforts must be focused on other, higher-priority issues. Incrementalism and resistance to change may be endemic to policymakers and bureaucracies.

The interaction that ensues from this conflict about the policy agenda has some beneficial consequences for both philosophy and policy. For policy, philosophers can contribute to a desirable widening of the policy agenda. Though philosophers may want to set agendas that are unrealistically wide, policymakers and bureaucrats may otherwise construe them unnecessarily narrowly. For philosophy, the result may be analyses of policy issues reflecting a more realistic understanding of the constraints of political reality. Nevertheless, philosophers' credibility problem can be exacerbated as they are seen as unrealistic, "head in the clouds," "ivory tower" academics. And more to my general point here, this mutual adjustment process can lead philosophers to a narrowing of vision and acceptance of fixed points that is contrary to the scholarly, philosophical virtues.

I have sketched a variety of respects in which the characteristic aims, virtues, and commitments of philosophical scholarship are in conflict with those of public policy-making. Does this imply that philosophers should avoid the policy process like the plague? I think not. Some aspects of the conflicts I have cited are beneficial either for policy and/or philosophy. In many respects that I have not touched on here the experience can lead to better applied ethics – perhaps most obviously in giving philosophers enough experience in a particular area of applied ethics to gain a clear understanding of its moral issues and problems. Far too often, philosophers fail to be as effective as they might be because they lack adequate sustained exposure and understanding of the area, such as medicine or business, in which they seek to do applied ethics. Moreover, there are many valuable roles, though I have not detailed them here, that I believe at least some philosophers are both professionally and personally well suited to play. Despite the philosophy-policy conflict I have been developing here, I believe that philosophers who are fortunate enough to have the opportunity to use their analytical and critical skills at influential points in the policy process often can help improve and illuminate thinking and practice in ways that produce real and significant benefits for those affected by the policies. I, at least, found that to be a deeply satisfying aspect of my own experience in the policy world. However, I believe the scholarly-policy conflicts that I have cited here do give reason for thinking that philosophers' forays into the world of policy should best be limited and temporary, not full time and permanent. The philosophical virtues that enable philosophers to make effective, valuable, and distinctive contributions to the policy process are probably best maintained if their primary base and commitment remain in academic philosophy.

NOTE

1 Norman Daniels, "Conflicting Objectives and the Priorities Problem," in *Income Support*, ed. P. Brown, C. Johnson, and P. Vernier (Totowa, N.J.: Rowman & Littlefield, 1981).

Index

417

425

Index

irrational decision response, 90–1
  in shared decision making, 69–70
  value neutrality claim, 70–7
plan of life, *see* life plan
pleasantly senile, 379–80
policy making, *see* public policy
positivism, 56–60, 77n2
preference theories
  correction of, 66, 271
  death risk reduction, 249–51
  and "good life," 65–6, 270–4
  in life prolongation programs, 240–1, 246–7, 249–51
  quality of life, 312–13
  and response to illness, 310–11
  self-determination in, 292–3
  subjectivity in, 272
  weighting approach, 273–4
premature death, 295–6, 396, 400
Presant, C. A., 318n40
President's Commission for Ethical Problems in Medicine, *Deciding to Forego Life-Sustaining Treatment*, 152, 153, 160, 169, 230n1, 315n17, 382n4, 383n7
President's Commission for Ethical Problems in Medicine, *Defining Death*, 145, 146, 382n5
President's Commission for Ethical Problems in Medicine, *Making Health Care Decisions*, 148, 149, 150, 315n14, 316n24, 353n11
President's Commission for Ethical Problems in Medicine, *Securing Access to Health Care*, 158, 353n15
President's Commission for the Study of Ethical Problems in Medicine and Biomedical and Behavioral Research, 21n, 53n3, 77n1, 91n1, 315n14
"prevention paradox," 90
primary functional capacities

adjustment to limitations, 306–8
quality of life measure, 348–9
self-determination component, 67–8, 311–14
and subjective satisfaction, 305–14, 349
types of, 67–8
probability, 85–6
"professional practice" standard, 49–50
prohibitions, 97
prolonging life, 235–64
  certain death versus risk of death, 249–51
  cost effectiveness measures, 251–64
  equal opportunity claim, 255–6
  human capital approach, 236–8
  individual values in, 245–51
  just distribution issues, 254–5, 263
  lost earnings approach, 236–8
  quality-adjusted life years measure, 251–64
  relative value of, 253–4
  value of, 235–64
  value to others approach, 260–1
  willingness to pay test, 238–51
prudential allocator approach
  and age-rationing, 16, 358–60
  elderly, 358–60, 391–8, 402–3
  and interpersonal distribution approach, 360
  life-extending care, elderly, 402–3
  moderate dementia, 379–81
  persistent vegetative states, 365–70
  personal identity assumptions, 382n3
  severe dementia, 376–8
"prudential lifespan account," 391
psychiatric evaluation, 226
psychological continuity identity theory, 366–9

Index

Rawls, J., 14, 17n5, 17n6, 63, 101, 119, 120n4, 121n10, 142n5, 288, 298, 317n33, 318n45, 352n4
"reasonable person" standard
life-sustaining treatment decisions, 128–9, 155–6
in risk disclosure, 50
Reece, W. S., 318n40
Rees, G. J. C., 352n5
reflective equilibrium
permissible killing decisions, 101
in treatment choices, 63–4
reimbursement to physicians, 328–9
relief of suffering, 173–5
religious beliefs
and euthanasia, 213
and irrational decisions, 87
renal dialysis, 331–2
Report of the Joint National Committee on Detection, Evaluation, and Treatment of High Blood Pressure, 318n40
Rescher, N., 265n10
resource allocation
certain death versus risks of death, 249–51
demented patients, 15–16
distributive justice theory, 115–20, 358–9
elderly, 16, 358–9, 388–416
health policy issues, 11–12
and permissible killing, 112–20
prudential approach, elderly, 358–60
willingness to pay methods, 12–13, 245–51
respirator support, 159–61
see also life support
revocable trust, 104
Rhoden, N. K., 283, 315n23
Richards, D., 121n10
Rie, M. A., 353n9
right of disposal, 385n17

right to life
conflict of interests, 117–20
distributive justice theory, 112–20
involuntary euthanasia, 109–11
patient-centered model, 118
relative strength of, 120
and voluntary euthanasia, 103–9
right to refuse treatment, *see* treatment refusal
right to self-determination, *see* self-determination
rights-based position
defense of, 6–7
versus Donagan's duty theory, 123–33
euthanasia application, 208–13
and permissible killing, 6, 97–120
role of autonomy, 127–33
seriously ill newborns, 177
and wishes of the victim, 125–8
risk disclosure, *see* disclosure of risks
risk evaluation, patients, 85–6
risks of death
versus certain death, 249–51
preferences in values, 250
quality-adjusted approach, 259
value to others approach, 261–2
and willingness to pay, 240–2, 249–51
Roman Catholic moral theology, 168
Rorty, A, 383n9
Rose, G., 92n13
Rosner, F., 157
Roth, L., 53n4
Rothman, D., 17n2
Ruark, J. E., 315n13

Salgo v. Leland Stanford Jr. University Board of Trustees, 53n1
Sayre-McCord, G., 62
Scanlon, T. M., 314n3

Index

Index